DATE DUE

NORTH AMERICAN FREE TRADE AGREEMENT

Also by Khosrow Fatemi

INTERNATIONAL TRADE: A North American Perspective
INTERNATIONAL TRADE: Existing Problems and Prospective
  Solutions
SELECTED READINGS IN INTERNATIONAL TRADE
THE MAQUILADORA INDUSTRY: Economic Solution or
  Problem?
U.S.–MEXICAN ECONOMIC RELATIONS: Problems and
  Perspectives

North American Free Trade Agreement

Opportunities and Challenges

Edited by

Khosrow Fatemi

Dean, Graduate School of International Trade
and Business Administration
Laredo State University, Laredo, Texas

St. Martin's Press

First published in Great Britain 1993 by
THE MACMILLAN PRESS LTD
Houndmills, Basingstoke, Hampshire RG21 2XS
and London
Companies and representatives
throughout the world

A catalogue record for this book is available
from the British Library.

ISBN 0-333-59588-2

Printed in Great Britain by
Antony Rowe Ltd
Chippenham, Wiltshire

\

First published in the United States of America 1993 by
Scholarly and Reference Division,
ST. MARTIN'S PRESS, INC.,
175 Fifth Avenue,
New York, N.Y. 10010

ISBN 0-312-09976-2

Library of Congress Cataloging-in-Publication Data
North American Free Trade Agreement : opportunities and challenges /
edited by Khosrow Fatemi.
p.   cm.
Includes index.
ISBN 0-312-09976-2
1. Free trade—North America.   2. North America—Commerce.
3. North America—Commercial policy.   4. Canada.   Treaties, etc.
1992 Oct. 7.   I. Fatemi, Khosrow.
HF3211.N666   1993
382'.911812—dc20   93-18906
                                                    CIP

Contents

Preface

The global trading system of recent decades has been dominated by multilateralism and a multilateral approach to international trade and financial issues. Its many shortcomings notwithstanding, this system has generally been successful in bringing about a freer international trading system by reducing tariffs and other trade barriers. The system, symbolized and operationalized by the General Agreement on Tariffs and Trade (GATT), is based on a global, rather than regional, liberalization of trade. For obvious pragmatic reasons, it also includes many provisions for regional free trade agreements and economic cooperation in general. It is under the exemption allowed by these provisions that many regional trading blocs, the latest of which is the North American Free Trade Agreement (NAFTA), have been negotiated.

In many ways, multilateralism was a product of post-Second World War bipolarity and it flourished in the Cold War era. Like other products of the period, multilateralism has come under intense pressure and scrutiny in recent years. But unlike other products of the Cold War era, it seems to have survived, albeit without much of its glory and eminence. GATT, a rare success story in the international economic arena for most of its existence, has become paralyzed in its most recent round of trade negotiations. The Uruguay Round of multilateral trade negotiations has been stalled for over a year, and as soon as one barrier is seemingly overcome, a new problem develops. Whether it is European obstinacy, American stubbornness or Japanese obduracy, the fact remains that the Uruguay Round, scheduled for completion in 1991, still has a long way to go, and more and more seems unlikely to ever reach a meaningful conclusion. Despite the potential for mutual benefits, it seems that progress in regional free trade agreements (FTAs) is generally at the expense of, and not in conjunction with, multilateral trade agreements.

Some might argue that, at least in theory, regional free trade agreements are discriminatory and in the long run counter-productive; and that NAFTA is no exception. Those who subscribe to this hypothesis argue that regional FTAs discriminate against non-members and therefore weaken the effects of market forces. Therefore, in the long run, their impact will be to reduce global productivity. Their appeal is that this reduction in global productivity generates what amounts to short-term regional benefits and few countries are willing to forgo such benefits, and thus the recent surge in such agreements.

Such theoretical arguments notwithstanding, and from the regional perspective, the implementation of NAFTA will result in the creation of the world's largest trading bloc. More significantly, because of the diversity of, and the economic disparity between, its members, NAFTA's potential for growth is much greater than other similar attempts, including the European Community. Furthermore, NAFTA has a much greater potential for future expansion than other regional FTAs. For example, it could serve as a base for a 'Hemispheric' FTA, which would increase its potential substantially.

ORGANIZATION OF THE BOOK

This volume consists of five parts and 23 chapters, including an epilogue. Its organization is designed is to provide a balanced view of a very broad and somewhat controversial issue. The volume begins with an introductory chapter by the editor followed by an overview of the forthcoming debate on NAFTA by Professor Sidney Weintraub of the LBJ School of Public Affairs at the University of Texas in Austin. Weintraub provides an in-depth analysis of the agreement itself and identifies three issues which in his opinion will dominate the debate. 'These are whether the agreement and its parallel negotiations go far enough in protecting the environment, whether the parallel-track understandings on workers' rights are adequate, and how best to deal with worker dislocation resulting from NAFTA even if, on the whole, the agreement helps to create jobs in the United States.'

In discussing the approach adopted by the negotiators, Professor Weintraub divides the text of the 2000-page *North American Free Trade Agreement* into four parts: (a) the core agreement of provisions applicable to all three countries; (b) the chapter annexes, which contain exceptions to the core understandings; (c) transitory reservations; and (d) the tariff-reduction schedules. The latter two are temporary provisions and after a period of transition only the first two parts will remain in force. Professor Weintraub, who gives credit for the initiation of NAFTA to the Mexican government, concludes by encouraging its ratification: 'Just as it would have been remarkable had the U.S. administration spurned the Mexican initiative to negotiate a free trade agreement, so will it be extraordinary if Congress rejects the agreement.'

Part II includes an introduction followed by a discussion of the macroeconomic issues involved in the negotiation and implementation of NAFTA. Contributors to this part are Dr Robert Shelburne, an economist with the U.S. Department of Labor, who in Chapter 4 makes a comparison between NAFTA, specifically the inclusion of Mexico in NAFTA, and the southern expansion of the European Community in the 1980s when Greece, Spain and Portugal were admitted. Shelburne's contribution is followed by Chapter 5, in which Professor M. Reza Vaghefi from the University of North Florida discusses NAFTA's potential as a base for a North American common market. In Chapter 6 Professors Barry W. Poulson and Mohan Penubarti from the University of Colorado at Boulder study the impact of the debt crisis of the 1980s on regional (North American) trade.

Part III consists of an introduction and six chapters dealing with national perspectives and bilateral issues. The Canadian perspective of NAFTA is the topic of Chapters 8 and 9. In these two chapters Professor Ricardo Grinspun from the University of York and Edward Bruning from the University of Manitoba present somewhat opposing views of the impact of NAFTA on Canada. In Chapter 8, Professor Grinspun analyzes some preliminary results of the Canada–U.S. Free Trade Agreement – and potentially of NAFTA – on the Canadian manufacturing sector. In Chapter 9, Professor Bruning presents a more optimistic picture of Canada–U.S. FTA and a more promising future for NAFTA.

The Mexican perspective of NAFTA is presented in Chapters 10 and 11. In these two chapters Professor Joseph McKinney from Baylor University and Professor Peter Morici from the University of Maine discuss the role of Mexico in NAFTA and the impact of the latter on the Mexican economy. Professor McKinney presents the potential benefits of NAFTA for all three member-states, while Professor Morici discusses the political and economic problems of not implementing the agreement.

The remaining two chapters of Part III are devoted to bilateral U.S.– Mexico economic relations. In Chapter 12, Professor Gorge G. Gonzalez from Trinity University and Professor Alejandro Velez from St Mary's University discuss intra-industry trade between Mexico and the United States. Following a similar pattern, in Chapter

13 Professors Kurt Jesswein from Laredo State University, Stephen Salter and L. Murphy Smith from Texas A&M University address the issue of foreign-exchange relations between the two countries.

Following an introduction, the six chapters in Part IV are devoted to industry-specific and cross-border issues. In Chapter 15, Dr James Lane from MAREA International examines the correlation between training and labor turnover among the maquiladora plants in the U.S.–Mexico border region. In Chapter 16, Professor Jane LeMaster from Laredo State University and Professor Bahman Ebrahimi from the University of North Texas use the internal rate of return method of analysis to study the significance of human capital investment in the same border region. Also dealing with the U.S.–Mexico border region is Chapter 17, in which Professor Lawrence W. Nowicki from Adelphi University discusses the findings of his study of maquiladora plants in Sonora, Mexico. The last three chapters of Part IV are studies of different industries and the potential impact of NAFTA on each industry in one or more countries. The fruit and vegetable industry is the subject of Chapter 18 by Professor Nancy Wainwright from Eastern Washington University. Chapters 19 and 20, on the other hand, present two studies of the textile and apparel industry. In Chapter 19, Professors Sandra Forsythe, Mary Barry and Carol Warfield from Auburn University analyze the impact of NAFTA on the international competitiveness of the member states, particularly the United States. In Chapter 20, Professors Kathleen Rees from University of Nebraska, Lincoln, Jan M. Hathcote from the University of Georgia and Carl L. Dyer from the University of Tennessee at Knoxville adopt a more micro approach to study the impact of NAFTA on specific sectors of the industry.

Part V contains an introduction and a single chapter devoted to environmental issues. In Chapter 22 Dr John Audley from the Sierra Club discusses those specific provisions of NAFTA which relate to environmental issues. The volume concludes with an epilogue, which presents a summary of the major NAFTA-related events that have occurred since the rest of the volume was written, specifically those which took place during the latter months of 1992 and the early months of 1993.

In conclusion, regardless of any theoretical and practical benefits or shortcomings that NAFTA may have, and despite some vociferous opposition by different groups in all three countries, an agreement has been negotiated, and has a reasonable chance of becoming reality. Furthermore, because of the substantial economic disparity which exists among its member states, the North American Free Trade Agreement is unlike all other attempts to create regional free trade agreements. In short NAFTA is sailing into uncharted waters. To understand its implications requires a thorough examination of all possible and probable outcomes through extensive research. It is hoped that the present volume will serve as a step in this direction.

Laredo, Texas   KHOSROW FATEMI

Acknowledgements

This volume contains, for the most part, revised and updated studies which were first presented at the second international conference of the International Trade and Finance Association (IT&FA) held during April 22–5 1992 in Laredo, Texas. The support of the officers and directors of the International Trade and Finance Association was instrumental in organizing the conference, and consequently in completing this volume. My special thanks go to the IT&FA's family, particularly its past presidents, Peter Gray and Franklin Root, and its 1992 president and president-elect, Robert Baldwin and Mordechai Kreinin.

My very sincere thanks also go to the authors of the papers first presented at the conference and now published in this volume. The timely submission of their contributions, and their observance of editorial instructions made my work much easier. In short, it was a real pleasure working with such professional colleagues.

The conference was a cooperative effort between IT&FA and Laredo State University. My special thanks go to the members of the University community for their support and invaluable assistance not only in the preparation of the conference, but also in editing this volume. In particular, I am grateful to Leo Sayavedra, the president of Laredo State University, for his commitment to the internationalization of business education at the University and for providing continuous encouragement and support for this volume.

Finally, the time and effort that Sue Nichols, my able assistant at *The International Trade Journal*, put into the preparation of this conference was far beyond the call of duty. I am grateful to her for her devotion to her work, which is exemplified by her work on this volume. She made my work easier and more fun.

Laredo, Texas   KHOSROW FATEMI

Notes on the Contributors

Khosrow Fatemi, the editor, is Dean of the Graduate School of International Trade and Business Administration at Laredo State University. He is Professor of International Trade at Laredo State University, Laredo, Texas where he teaches graduate courses in international trade and international business. Dr Fatemi gained a PhD in International Relations and an MBA at the University of Southern California. His academic experience also includes university teaching in Europe, the Middle East and the Far East. Dr Fatemi is the founding editor of *The International Trade Journal*. He has published more than twenty books and articles and has presented a large number of papers at different professional meetings around the world. He is the editor-in-chief of the International Business and Trade Series for Taylor and Francis International Publishers, and as a founding member of the International Trade and Finance Association is presently serving as executive vice-president.

John J. Audley is a Research Fellow at the Sierra Club Center for Environmental Innovation. His focus is trade analysis for the Sierra Club's international program. Dr Audley gained a Master of International Business Management degree at the American Graduate School of International Management and is a PhD candidate at the University of Maryland, College Park.

Mary E. Barry is Associate Professor in the Department of Consumer Affairs at Auburn University. She received her EdD in vocation education from Temple University. Dr Barry's research interests include worldwide production and distribution of textiles and apparel; global retailing and sourcing; and apparel apprenticeship programs.

Edward R. Bruning is Professor of Marketing and Director of the Centre for International Business Studies at the University of Manitoba. He received his PhD in Economics from the University of Alabama. Dr Bruning has served as consultant to the U.S. Department of Transportation, the State of Ohio, and various municipal and county governments. In addition, he has assisted numerous businesses in matters concerning international business, exporting, transportation, marketing and the environment. He has published over 60 academic articles and proceedings.

Carl L. Dyer is Associate Professor of Retail and Consumer Sciences at the University of Tennessee in Knoxville. His research interests include international business and strategy, with a special interest in international retailing.

Bahman Ebrahimi is Associate Professor of Strategic Management and International Business and the Chair of the College of Business Administration Committee on International Business at the University of North Texas. He received his PhD from Georgia State University. Dr Erbahimi co-authored *International Business* and has authored or co-authored more than fifty papers and articles in professional journals and proceedings.

Sandra Forsythe is Wrangler Professor in the Department of Consumer Affairs at Auburn University. She received her PhD in Consumer Economics and Marketing from the University of Tennessee at Knoxville. Dr Forsythe's research areas include apparel marketing, consumer behavior and international marketing. She has published in a number of journals on clothing and textiles, applied psychology and business and retailing.

Jorge G. Gonzales is Assistant Professor of Economics at Trinity University. His areas of research include international trade between Mexico and the United States, illegal immigration and international factor mobility. He has had articles published in the *International Economic Journal*, the *Journal of Business and Economic Perspectives* and *Southwest Business Review*.

Ricardo Grinspun teaches economics and is a fellow of the Centre for Research on Latin America and the Caribbean (CERLAC) at York University, Toronto. He received his PhD in Economics from the University of Michigan, Ann Arbor. Dr Grinspun directed at CERLAC, a collaborative research project on NAFTA with the participation of about 30 scholars from Mexico, the United States and Canada. He is co-editor and co-author of *The Political Economy of the North American Free Trade*.

Jan M. Hathcote is Assistant Professor of Textiles, Merchandising and Interiors at the University of Georgia in Athens. Her research interests include retailing and international trade of textiles and apparel.

Kurt R. Jesswein is Assistant Professor of International Finance and Banking at Laredo State University. He received his PhD in International Business from the University of South Carolina.

James M. Lane is the Director of Research for MAREA International, a marketing and management research firm which carries out extensive work in Mexico. MAREA conducts competitive intelligence projects and corporate relocation studies in addition to maquiladora-related research. He received his DBA in International Business from United States International University. Dr Lane is an adjunct faculty member at several San Diego area universities.

Jane LeMaster is Assistant Professor at Laredo State University. She received her PhD in Organization, Theory and Policy from the University of North Texas. Dr LeMaster is the author of numerous articles on international management and international technology transfer with a special interest in Mexico.

Joseph A. McKinney is Ben H. Williams Professor of Economics at Baylor University. He received his PhD in Economics from Michigan State University. He has taught at the University of Virginia, and as an exchange professor at Seinan Gakuin University in Fukuoka, Japan. Dr McKinney is the co-editor of *Region North America: Canada, The United States and Mexico* and *A North American Free Trade Region: Multidisciplinary Perspectives*, and has authored several articles on different aspects of North American free trade. He co-directed a study on the subject for the Joint Economic Committee of the U.S Congress, and has testified before government committees and agencies on the anticipated economic effects of NAFTA.

Peter Morici is a Professor of Economics and Director of the Canadian–American Center at the University of Maine. He received his PhD from the State University of New York at Albany. Dr Morici has served at the National Planning Association, the Iacocca Institute, the Atlantic Council, the Congressional Office of Technology Assessment Advisory Group (trade with Mexico), the North American Economics and Finance Association. He has directed the Council on

Foreign Relations study group on the future of the Canada–U.S. Free Trade Agreement. Dr Morici is the author of many monograph and journal articles as well as eight books, including *Trade Talks with Mexico: A Time For Realism* and *Making Free Trade Work: The Canada–U.S. Agreement.*

Lawrence W. Nowicki is Assistant Professor of International Business and Finance at Adelphi University in New York. He received his PhD from the University of Paris. He has taught at the American Graduate School of International Management and the University of Lyons. Dr Nowicki has been a consultant for the OECD, the World Bank, the U.N. Centre for Transnational Organizations and several European and U.S.-based corporations. He has conducted research for the French National Center for Scientific Research and the Flagstaff Institute.

Mohan Penubarti is a doctoral student in Economics and in International Relations at the University of Colorado. His research interests are in the area of international economic competition, particularly in high-technology industries. He is a graduate of the University of Michigan College of Engineering.

Barry W. Poulson is Professor of Economics at the University of Colorado. He serves as Adjunct Scholar with the Heritage Foundation, and is a Senior Fellow of the Independence Institute. Dr Poulson has written numerous books and articles in the field of economic growth and development, and constitutional economics. He is currently co-editor of *The North American Review of Economics and Finance*. He is past president of the North America Economics and Finance Association.

Kathleen Rees is Assistant Professor of Textiles, Clothing, and Design at the University of Nebraska in Lincoln. She is a PhD candidate in Consumer Environments at the University of Tennessee in Knoxville.

Stephen B. Salter is Associate Professor of International Accounting at Texas A&M University. He received his PhD from the University of South Carolina. Dr Salter has published in a variety of journals, including *Advances in International Accounting* and *The Journal of International Business Studies*. He was previously a partner in Ernst and Young Consulting and has worked for a number of international banks in Asia, Canada and the Caribbean.

Robert C. Shelburne is a Senior International Economist in the Division of Foreign Economic Research of the United States Department of Labor. He received his PhD in International Economics from the University of North Carolina at Chapel Hill. Dr Shelburne has taught at Ohio University and North Carolina State University.

L. Murphy Smith is Associate Professor of Accounting at Texas A&M University. He received his DBA from Louisiana Tech University. He is a member of the American Institute of CPAs and the American Accounting Association. Dr Smith is co-author of *Trap Doors and Trojan Horses* and has published over 170 professional journal articles, research grants, books and professional meeting presentations. He is also microcomputer department co-editor of *The CPA Journal*.

M. Reza Vaghefi is Professor of Business Administration and Director of the International Business Program at the University of North Florida's College of Business Administration. He received his PhD from Michigan State University. Dr Vaghefi has also taught at the American University of Beirut, Tehran University, Concordia University in Canada and Michigan State University. He has written five books (in two languages) in business and economics, articles in international journals and numerous case studies.

Alejandro Velez is Professor of Economics at St Mary's University, San Antonio. His area of specialty includes international economics, Latin American economics and macroeconomics. He has had articles published in the *Journal of Economics and Business Research*, *Southwest Journal of Business and Economics*, the *Texas Business Review*, *El Financiero* and *Business and Economic Dimensions*.

Nancy A. Wainwright is Assistant Professor of Business Law at Eastern Washington University. She was the College of Business Administration's Outstanding Teacher for 1991–2.

Carol Warfield is Professor and Head of the Consumer Affairs Department at Auburn University. She received her PhD in Family and Consumption Economics from the University of Illinois. Dr Warfield's research is in worldwide production and distribution of textiles and apparel, textile and apparel industry competitiveness, and consumer-wear studies.

Sidney Weintraub is Dean Rusk Professor of International Affairs at the Lyndon B. Johnson School of Public Affairs at the University of Texas at Austin. He is also a distinguished scholar at the Center for Strategic and International Studies in Washington, DC. He received his PhD in Economics from the American University. Dr Weintraub serves as an economic consultant to the U.S. and Mexican governments, private corporations and consulting firms as well as the International Monetary Fund, the World Bank, InterAmerican Development Bank and the United Nations. He is the author of *A Marriage of Convenience: Relations between Mexico and the United States* and is editor of *U.S. Mexican Industrial Integration: The Road to Free Trade.*

Part I
Overview

1 Introduction

Khosrow Fatemi

1.1 HISTORICAL PERSPECTIVE

The origin of the North American Free Trade Agreement (NAFTA) dates back to 1988 when President Ronald Reagan of the United States and Prime Minister Brian Mulroney of Canada initiated the Canada–U.S. Free Trade Agreement (Canada–U.S. FTA). To some extent the Reagan–Mulroney decision was an extension of their philosophical beliefs in free trade and limited government intervention in economic affairs. Both Ronald Reagan and Brian Mulroney are ardent conservatives with strong beliefs in a *laissez-faire* approach to economics. Additionally, both had been elected on conservative platforms and were anxious to implement their respective campaign promises. Finally, the economies of the United States and Canada were so closely interwoven that Canada–U.S. FTA was a natural extension of existing activities. In fact, it can be argued that Canada–U.S. FTA was a mere institutionalization of a *de facto* economic union.*

The 1986 Reagan–Mulroney decision to start negotiations to form the Canada–U.S. FTA was not the first integration attempt between the two countries. The history of such attempts date back to the mid-1800s.† Historically, Canada has been more reluctant to seriously consider an economic union with the United States. Economically, this

---

* This is not to say that economic differences did not exist. Indeed there were – and still are – serious economic and trade disagreements between the two countries. Recent examples include the arguments over the lumber industry and the Honda case. However, such isolated cases notwithstanding, economic ties between Canada and the United States in the mid-1980s were as close as they are between any two countries without a formal integrative agreement.
† The 1854 negotiations are generally recognized as the first successful attempt at developing an economic union between the United States and Canada. The 1854 negotiations actually resulted in the signing of a free trade agreement between the two countries. The agreement included only natural resources and agricultural products, but not manufactured goods. The agreement was abrogated in 1866 by the United States after repeated attempts by both countries to expand it into a customs union failed, and in part in response to political problems stemming from the American Civil War. For an excellent description of this agreement see Harold Crookell, *Canadian–American Trade and Investment Under the Free Trade Agreement* (New York: Quorum Books, 1990) pp. 5–7.

3

is the perfect manifestation of fear of survival and is the typical reaction of the smaller partner in any integration attempt. In other words, Canadians, particularly Canadian businesses that feared being absorbed by their much larger U.S. counterparts, opposed the Canada–U.S. FTA. Canadian businesses also felt that because of their smaller size, any expansion of trade with the United States would make them too dependent on one country. After all, even before Canada–U.S. FTA (1988), more than 70 per cent of Canada's exports, were destined for the United States, while only 21.8 per cent of America's exports were to Canada (see Table 1.1). They felt a free trade agreement would increase this 'lopsided interdependency' and make the Canadian economy too dependent on the United States.

The significance of economic issues notwithstanding, Canada's traditional reluctance for a formalized union with the United States has been socio-cultural and political. Canadians, at least a large percentage of them, contrast the United States' 'melting pot' with Canada's 'mosaic'.[1]

Table 1.1   Selected economic indicators, Canada and the United States before CUSFTA, 1988

	Canada	U.S.	Canada as % of U.S.
Population	25.9	245.1	10.6
Gross national product	586.2	4908.0	11.9
Total exports	116.4	319.4	36.4
Exports to the other country	82.0	69.7	—
Bilateral exports as % of total exports	70.4	21.8	—
Total imports	110.1	459.8	23.9
Imports from the other country	70.5	81.4	—
Bilateral imports as % of total imports	64.0	17.7	—
Jobs created by exports to the other country* (millions of jobs)	1.64	1.39	—

* Based on the assumption that each $1.0 billion of exports create 20000 jobs.
Source: *Direction of Trade Statistics Yearbook 1992*, and *International Financial Statistics 1992*.

1.2   THE ROLE OF 'EUROPE-1992'

In the Fall of 1985, the Council of Ministers of the European Community, meeting in Luxembourg, drafted the Single European

Act (SEA). The Act, which actually went into effect on 1 July 1987, had the purpose of creating a single (internal) market, defined as 'an area without internal frontiers in which the free movement of goods, persons, services and capital is ensured'.[2] The fact that a specific deadline (the end of 1992) was set for complete implementation of the different provisions made the SEA much more serious – and from the outsiders' point of a view much more ominous – than previous pronouncements of the European Community. To many non-Europeans, Europe-1992 sounded too much like 'Fortress Europe' in which rapid expansion of intra-Community trade would come primarily at the expense of the rest of the world.

The prospects of losing free access to the European market were rather discouraging for both Canada and the United States. Both countries have traditionally relied on Europe as a major trading partner as well as a reliable investment partner (see Table 1.2). Both countries felt threatened by the prospect of Europe-1992, and neither wanted to be unprepared when 'Fortress Europe' became a reality.

Table 1.2   Trade relations with the European Community, 1990

	Canada	U.S.
Exports to EC Countries ($bn)	9.9	103.1
Total exports ($bn)	127.2	422.2
EC as % of total exports	7.8	24.4
Imports from EC ($bn)	12.6	89.4
Total imports ($bn)	121.7	509.0
EC as % of total imports	10.4	17.6

Source: *Direction of Trade Statistics Yearbook*, 1991, pp. 123–5, 402–4 (percentages calculated by the author).

Faced with the prospect of facing new trade barriers in Europe, both Canada and the United States began looking at other alternatives. This was particularly true for Canada.* In reality, Europe-1992 seems to

---

* In the extreme – albeit highly unlikely – case that Fortress Europe were to replace all of Canada's trade with intra-Community trade, Canada would lose 7.6 per cent of its total exports. This would also mean a loss of 200000 jobs in Canada. (Comparable numbers for the United States would be 24.9 per cent and almost two million jobs.) Even through this is an extremely unlikely result of Europe-1992, one could not escape the conclusion that even one-quarter of this would be devastating to Canada's economic prospects. Consequently, no Canadian – nor American, for that matter – leader would want to be caught unprepared for such an eventuality.

have acted as the final impetus needed to bring about serious negotiations for a Canada–U.S. Free Trade Agreement. However, many other factors also contributed to the formation of Canada–U.S. FTA. A partial list of these factors would include (a) the rapidly expanding cumulative trade deficit in the United States, (b) an accelerating trend around the world toward greater globalization of international business, (c) the narrowing gap in the United States' technological superiority and (d) a general decline of nationalism in the Western world. Given the above, the prospects of an emergence of protectionism in Europe served as the catalyst to start, and successfully complete, the free trade negotiations.

The negotiations started in 1986 and 18 months later concluded with the signing of the Canada–U.S. Free Trade Agreement. The agreement, which went into effect on 1 January 1989, created the world's second largest trading bloc, after the European Community. It should be noted that the similarities between the Canada–U.S. FTA and the SEA end here. The objectives of the former are limited to the elimination of trade barriers between the two countries. The SEA, on the other hand, is a very ambitious and comprehensive treaty aimed at the creation of a truly single market in which not only goods and services, but also labor and capital move freely.

1.3 ENTER MEXICO

Mexico's international economic policy has always been dictated, in varying degrees, by a combination of several, often contradictory, goals and concerns. Even though the relative significance of these objectives has fluctuated over the years, the dominant consideration has always been to create a balance between economic ties with the United States while at the same time reducing the political reliance on the United States – or the perception thereof. In the international political arena, Mexico has always had ambitions of becoming a leader of the developing countries and this goal cannot be accomplished if the country is too closely affiliated with 'Yankee imperialism'. Historically, Mexican nationalism has not forgotten – much less forgiven – Mexico's wars with, and the loss of large territories to, the United States.

Such feelings began to moderate in 1982 when Miguel de la Madrid, a trained economist and a firm believer in an open economy and free trade, was elected President of Mexico. In the earlier years of his presidency, he faced major economic problems best illustrated by the loss of 83.3 per cent in the value of the Mexican peso in a single year (1982).* Once the economic crisis in Mexico had been brought under relative control, the government began reversing – albeit very slowly – the protectionist policies of the earlier years by taking measures to open the Mexican economy to free trade. Mexico's accession to GATT in 1986 was a major step in this direction.

The pace of implementing free trade policies accelerated with the election of Carlos Salinas de Gortari to the presidency of Mexico in 1988. President Salinas, also a trained economist, initiated one of the most ambitious and comprehensive privatization efforts in the developing world. His philosophical orientation towards free trade notwithstanding, President Salinas found Mexico's heavy dependence on the U.S. market alarming. In 1987, the last full year of de la Madrid's presidency, almost three-quarters of Mexico's exports were destined for the United States, and this was not an atypical year (see Table 1.3).

Table 1.3   Selected economic indicators, Mexico and the United States, 1991

	Mexico	U.S.	Mexico as % of U.S.
Population (mid-year)	87.8	252.7	34.7
Gross national product ($ bil.)	240.0	5685.8	4.2
Total exports ($ bil.)	38.9	422.2	9.2
Exports to the other country ($ bil.)	29.0	33.3	—
Bilateral exports as % of total exports	74.6	7.9	—
Total imports ($ bil.)	22.1	509.0	4.3
Imports from the other country	15.6	31.9	—
Bilateral imports as % of total imports	70.6	6.3	—
Jobs created by exports to the other country* (000 of jobs)	580	666	—

*Based on the assumption that every $1.0 in exports will result in 20000 jobs.
Sources: International Financial Statistics Year 1992, pp. 504–8, 716–21; and *Direction of Trade Statistics Yearbook 1992*, pp. 279 and 402–3 (percentages calculated by the author).

---

* The Mexican peso was devalued from 12.5 per dollar to 25 per dollar in 1976. Despite adverse economic charges, including high inflation, successive Mexican administrations maintained the 1976 value of the peso until February 1982. When the peso was devalued to 50 per dollar in February 1982, it was simply too little and too late. By the end of 1982, the peso's value had diminished to 150 per dollar.

In early Spring 1990, President Salinas made a much-publicized trip to Europe. The stated objective for the trip was to expand economic ties between Mexico and Europe, that is, to bring more European investment and trade to Mexico. By this time European lenders were fully preoccupied with the events of Eastern Europe and the Soviet Union. At every stop President Salinas was reminded of the economic burden that the events in Eastern Europe were placing on the 'limited' resources of Western Europe. In other words, the message that the European leaders conveyed to Salinas was that any solution to Mexico's economic woes, at least in the short run, must come from Mexico's own region, that is, the United States. Shortly after his return from Europe, President Salinas met with President Bush and the two agreed to start preliminary negotiations for a free trade agreement between the two countries.

The main approach to the bilateral negotiations between the United States and Mexico was to duplicate the Canada–U.S. Free Trade Agreement. From the beginning Canada was given the option of joining the negotiations as a full partner. When Canada chose to exercise this option in 1991, the nature and context of the goal of the negotiations changed from attempting to establish a bilateral U.S.–Mexico free trade agreement to trilateral negotiations for the establishment of a North American Free Trade Agreement.

1.4 NAFTA: GOALS AND OBJECTIVES

Chapter One, Article 102 of the agreement defines the objectives of NAFTA. Specifically, according to this Article, 'the objectives of this Agreement, as elaborated more specifically through its principles and rules, including national treatment, most-favored-nation treatment and transparency are to:

(a) eliminate barriers to trade in, and facilitate the cross border movement of, goods and services between the territories of the Parties;
(b) promote conditions of fair competition in the free trade area;
(c) increase substantially investment opportunities in their territories;
(d) provide adequate and effective protection and enforcement of intellectual property rights in each Party's territory;
(e) create effective procedures for the implementation and application of this Agreement, and for its joint administration and the resolution of disputes; and

(f) establish a framework for further trilateral, regional and multi-lateral cooperation to expand and enhance the benefits of this Agreement.'[3]

Two sections of this Article are particularly noteworthy. First, according to section (c), the three countries commit themselves to 'increase substantially investment opportunities in their territories'. This section is a departure from other free trade agreements in that issues related to investment are generally considered to be part of free trade agreements. In effect, this section goes one step beyond this provision, which is generally not included in free trade agreements. Additionally, this clause is an obvious concession by the Mexican government, as the other two countries have few limits on investment opportunities. Implementation of this section will require Mexico to revise drastically many of its investment laws, particularly those which impose maximum limits on investment by foreign firms and/or in certain industries. Other provisions and annexes of the agreement specify time-tables for the implementation of this section. They also include maximum limits for certain industries.

Section (f) of this article is also interesting and noteworthy. It commits the member states 'to establish a framework for further trilateral . . . cooperation to expand and enhance the benefits of this Agreement'. This section opens the door to two different possibilities. First, 'further trilateral cooperation' can mean expansion of NAFTA into a more comprehensive integration, such as a customs union. This would indeed be a departure from what has officially been discussed. It would also be a natural follow-up to what has already been accomplished. Currently, Canada and the United States have similar tariff rates, while Mexico's rates – already reduced substantially by virtue of its entry into GATT – are still much higher. One of the problems of implementing a free trade agreement is the potential abuse by importers who would reroute their products to countries with lower tariffs.*

---

* For example, currently Mexican has relatively high tariff rates on imported electronic products. After NAFTA goes into effect, the rates will be reduced and eventually eliminated for U.S. or Canadian-made electronic products, but will remain in effect for the same products imported from other countries. A Mexican importer of, say, Japanese VCRs could then reroute his imports through the United States, repackage them and import them into Mexico duty free. Rules-of-origin clauses of the agreement are supposed to stop such abuses, but they also require extensive monitoring of the border between the three countries. Such activities would not be conducive to liberalized trade, as critics could argue that tariffs have been replaced with administrative barriers.

1.5   NAFTA vs GATT: SOME OBSERVATIONS

Since the end of the Second World War, the dominant theme of international economic cooperation has been multilateralism. By definition, multilateralism encourages global agreements and discourages regionalism.* Yet, and despite the obvious success of the International Monetary Fund, the World Bank and the General Agreement on Tariffs and Trade – all global institutions established during the 1940s – recent years have witnessed a growing tendency on the part of many nations to form regional economic groupings.[†] The primary example of successful regionalism is, of course, the European Community, which started with six countries (Belgium, France, Germany, Italy, Luxembourg and the Netherlands) in 1960; expanded to nine in 1973 with the addition of Britain, Denmark and Ireland in 1973; then to ten in 1981 (with the accession of Greece) and twelve in 1986, when Spain and Portugal joined. The European Community is expected to include at least 20 countries by the end of the decade. Many other regional groupings, usually among the developing countries, have been attempted and have usually failed.

Whether regionalism supports or threatens multilateralism, and whether it matters if it does, there is one aspect of the recent trend that deserves some attention here. It is not too farfetched to argue that the relative global peace of recent decades is at least in part due to the 'global interdependence' brought about by multilateralism in trade and financial matters. Philosophically, and in practice, regionalism changes global interdependence to regional interdependence. Less global interdependence, at least theoretically, reduces the 'peace incentive', that is, the incentive for nations to respect each other's integrity and sovereignty.

---

* GATT, the primary vehicle of multilateralism in international trade, does indeed have some provisions (Article XXIV) to facilitate those countries which want to form regional groupings. Philosophically however, GATT symbolizes multilateralism and would lose its *raison d'être* if regionalism were to dominate the international economic arena.
[†] Some might argue that regionalism is spreading not despite but rather because of the success of such institutions as GATT. This argument, at best limited to trade issues, maintains that GATT has achieved just about as much as it will ever achieve. Average global tariff rates are now a small fraction of what they were when GATT was initiated. (It is true that non-tariff-barriers have in many instances replaced tariffs, but even so, trade is freer today than at any time in history.) The only way that further economic cooperation – and this is the core of this argument – can be achieved is by limiting the number of participants. Simply put, logistically it is not realistic, for example, to attempt to negotiate a global common market.

1.6 FUTURE PROSPECTS

Free trade agreements are not a new concept. They have been attempted many times before. What distinguishes NAFTA from all previous attempts, in addition to the $6 trillion size of the region's economy, is the wide disparity between the members' economic status. The three countries are simply not in the same economic stratosphere. The United States, the world's largest economic power, has an annual GNP of $5.7 trillion per year and a per capita income of $21 790. Canada, much smaller in total size ($586 billion) has a similar per capita income ($20 470). Mexico, on the other hand, has a per capita income of $2490. This is one-ninth of the comparative figure for the United States or Canada.* Table 1.4 compares the economies of the three NAFTA members.

Table 1.4 Comparison of selected economic variables, Canada, Mexico and U.S.

	Canada	Mexico	U.S.
Per capita GNP (1990 $)	20470	2490	21790
Exports as % of GNP	20.0	16.2	6.9
Literacy rate (all adults, %)	>95.0	>95.0	87.0
Average annual population growth (1980–90, %)	1.0	2.0	0.9

Source: *World Development Report 1992*, p. 219.

1.6.1 Short-term Prospects

The short-term impact of these economic disparities is to make both the negotiations and the implementation of NAFTA more difficult. Specific areas of difficulty include the following.

1.6.1.1 Infrastructure

Mexico's lagging infrastructure development will serve as a major impediment to a rapid expansion of trade among the three countries.

---

* As measured by per capita income, prior to NAFTA, the largest gap between two partners in an economic union of any type was less than five to one. In that case, Germany's per capita GNP of $22320, the highest in the European Community, was 4.7 times that of Portugal's $4900, the lowest in the Community.

The most direct land road between the three countries stretches from Mexico City northward into the United States and continues to the Canadian border.* Yet, only a small fraction of this road in Mexico is divided highway and the rest is not equipped for heavy trucks or for fast-moving passenger vehicles.[†] Additionally, road services such as restaurants and gasoline stations are infrequent and inadequate.[4]

1.6.1.2 Labor and Wages

Probably the most researched – and controversial – aspect of the economic gap between Mexico and the other two NAFTA partners is in the area of labor and wages (see Table 1.5).

Table 1.5   Selected labor statistics, Canada, Mexico and U.S.

	Canada	*Mexico*	*U.S.*
Total labor force (millions)	13.8	27.7	125.2
Unemployment rate (%)	10.3	3.5	6.8
Minimum wage ($ per hour)	4.00	0.55	4.25
Urban population as % of total	77	73	75

Source: *Statistical Abstract of the United States, 1991* (Washington, DC: U.S. Government Printing Office, 1991) pp. 384, 402; *World Development Report, 1992*, p. 279.

Even through free movement of labor is neither part of the FTA concept, nor is it expected to be included in NAFTA, three factors point to accelerated labor migration once NAFTA is implemented and some border barriers are removed. First, there are an estimated 2.10 million undocumented Mexican nationals living, and presumably 'working' in the United States.[§] Even if not intended, free trade between two neighboring countries makes border crossing 'easier'

---

* The North American section of this north–south highway starts at the Mexico–Guatemala border as Mexican Highway 190, continues northward through Mexico City and Monterrey and crosses the U.S. border at Laredo, Texas. It then continues as IH-35 north to Duluth, Minnesota, at which point it becomes U.S. Highway 61. Finally, it crosses the Canadian border at Grand Portage and continues as Canadian Highway 61.
[†] The only section of this road in Mexico which is divided highway is the 100-mile stretch north of Monterrey. Construction is currently underway to upgrade the remaining section of the Monterrey–Laredo road into a freeway.
[§] The range of this estimate is unusually wide because there are no check-points at which these immigrants are counted. The Immigration and Naturalization Service estimates are

and therefore more likely. The historical trend of northward migration is unlikely to reverse itself when barriers to this migration are actually reduced.* Secondly, NAFTA will bring more awareness to the masses in Mexico as to the extent of economic disparity between Mexico, Canada and the United States. This will be particularly important when a typical Mexican villager – who is now incognizant of the fact – finds out that he can increase his income tenfold by moving north. The appeal of such wealth can be nothing but overwhelming. Migration is not easy. It takes courage, endurance and just plain 'guts', but the lure should not be underestimated.

Economic theorists argue that NAFTA will eventually 'equalize' wage rates in the three countries.[†] This may be true in the long run. In this case however, 'long run' will be a very long run.[§] At any rate, it

---

in sharp contrast with some private estimates and all are inaccurate and highly simplistic. A survey of 656 Mexican nationals apprehended by the INS revealed that some individuals had previously crossed the border more than 40 times. During this study, the author observed numerous individuals who had been caught – not merely crossed – more than once in a single day. Such occurrences render estimates of the number of undocumented Mexican nationals in the United States highly imprecise. For a detail analysis of the results of this study, see Khosrow Fatemi, 'The Undocumented Immigrant: A Socioeconomic Profile', *Journal of Borderland Studies*, vol. II, no. 2, Fall 1987, pp. 85–99.

* Many, including Mexican President Salinas argue, correctly, that Mexicans migrate north in search of jobs and economic opportunity. 'Take the job to them and they will have no incentives to move', the argument goes. The premise of this argument is indisputable, but its conclusion is not. It is true that typically Mexican migrants move to the United States in search of better economic conditions. It is also true that many – maybe even most – of these people would lose their present incentive to migrate under NAFTA. However, improved economic conditions in Mexico will inevitably bring increased expectations for higher standards of living – not only higher but also higher than realistically possible. There are very few instances in the history of developing countries where improved economic conditions did not create expectations far beyond the means of the country. The recent uprisings in China, Hong Kong, Iran and Thailand, all of which followed periods of long and rapid economic growth, are all examples of this point. Many observers argue that these uprisings took place because respective governments failed to match their economic growth with political development. Maybe so. But one should not let the euphoria of 'global democratization' obscure the reality of many of these countries.

† According to the factor equalization theorem, barring barriers, factors of production, including labor, will cost the same in different countries. Whereas there is sufficient theoretical support for this argument, its application to NAFTA is faulty. After all, NAFTA is only a free trade agreement and not a common market. Even in the case of a common market and complete absence of any artificial barriers to the free flow of labor, one cannot disregard social and cultural barriers to such movements.

§ For example, given that current U.S. wages are approximately 10 times Mexican wages and assuming 3 per cent and 8 per cent per year increases for U.S. and Mexican wages respectively, it will take about 50 years for the two wages to become equal.

should be remembered that NAFTA, at least as envisaged now, does not include free movement of labor, and barriers will increase the length of time for attaining this utopian goal.

1.6.1.3 Financial Services and Banking

Mexican interest rates have been falling rapidly in recent years. Still, during 1991 peso-denominated money market and treasury bill rates in Mexico averaged 23.6 per cent and 19.3 per cent respectively. Comparable dollar-denominated rates in the United States were about 5.6 per cent. Furthermore, recently the peso has been quite stable – it lost only 4.3 per cent of its value during 1991. It is very safe to assume that Mexico, anxious to attract foreign investment and cut its inflation, is highly unlikely to accelerate the peso's devaluation rate. The dilemma for a Mexican national with some liquidity is whether to (a) invest his pesos in Mexico and earn between 19.3 and 23.6 per cent per year, or (b) convert his pesos into dollars and earn 5.6 per cent, then convert his dollars back into pesos and earn another 4.3 per cent because of the dollar's appreciation against the Mexican peso. The latter is not much of a choice – under 10 per cent in the United States vs over 20 per cent in Mexico – yet millions of dollars of Mexican funds are regularly deposited in U.S. banks or CDs.

Interest rates are not the only financial indicators showing major differences between Canada and the United States on the one hand and Mexico on the other. Table 1.6 summarizes the major financial variables, all of which point to major differences between the three countries.

Table 1.6   Selected financial statistics, Canada, Mexico and U.S., 1991

	Canada	*Mexico*	*U.S.*
Interest rates (money market rates, %)	7.4	23.6	5.7
Inflation rate (%)	5.6	22.7	4.3
CPI (1985 = 100)	131.4	1723.8	126.6
International liquidity/total reserves minus gold ($ bil.)	16.3	17.7	66.7

Sources: *International Financial Statistics*, June 1992, pp. 134, 354 and 538.

1.6.2 Long-term Prospects

The economic disparity between Canada, the United States and Mexico is a major short-term obstacle to free trade. Government policy makers and corporate decision makers, trying to work within the new parameters of North America, will face this problem over and over again. However, once short-term barriers are overcome, the present disparity could actually be the greatest inducement for the success of NAFTA. Mexico – in overcoming its current and short-term trade problems, as NAFTA is expected to enable it to do – will become a quasi-unlimited market for Canadian and American products, services and capital. In turn, the United States and Canada – once they have overcome the initial problems of integrating their trade with Mexico – will also become quasi-unlimited markets for Mexican light industrial products and labor-intensive products in general.

1.6.3 Increased Mexican Exports

The United States and Canada import billions of dollars of labor-intensive electronic products, mostly from developing countries in the Far East. Proponents of NAFTA argue that most, if not all, of this trade should – and will – be shifted to Mexico. This proposition makes political, social and even economic sense.

Politically, a successful NAFTA will make Mexico stable and prosperous, a distinct advantage even in the post-cold-war era. Sociologically and economically, NAFTA can curb the northward flow of Mexican migrants by creating more employment opportunities for Mexican workers. Also economically, Mexico's proximity to the markets of the United States and Canada and the resultant reduced transportation and inventory costs should more than offset any benefits of lower Asian wage rates.

1.6.4 Increased Mexican Imports

Currently, Mexico's total imports amount to 9.2 per cent of its GNP. Furthermore, 81 per cent of Mexican imports are from Canada and the United States. NAFTA's implementation is expected to increase both of these percentages – albeit slowly. Assuming a 10 per cent annual increase in Mexico's imports from its North American neighbors, by the year 2000 Mexico will import $55.5 billion from Canada and the

United States. This translates into 638 000 new jobs in the two
countries.*

1.7 NAFTA: THE FIRST STEP TOWARDS A HEMISPHERIC FTA

In 1990, President Bush unveiled his 'Enterprise for the Americas
Initiative', calling for greater cooperation among the nations of the
Western Hemisphere. Even though a real 'Western hemispheric free
trade area' is many years away, and even though NAFTA is itself not
yet a reality, steps are already being taken for its future expansion.†
Because of the economic slowdown in Canada and the United States,
coupled with pending elections in both countries – and Mexico –
expansion talks are given low priority at this stage. Even without its
architect, George Bush's Enterprise for the Americas is likely to receive
serious consideration as a viable economic policy in the coming years.
On possible approach for implementing the Enterprise for the
Americas Initiative may be geographical expansion of NAFTA.
Should this approach be adopted, the first candidates are likely to be
Chile and Venezuela, followed by Colombia and Argentina.§ Some
Central America countries – because of their small size – can be

---

* In 1991 Mexico imported $23.6 billion from Canada and the United States. If increased
by 10 per cent per year for the rest of the decade, the comparable figure in the year 2000
would be $55.5 billion. Assuming that each $1.0 billion of exports will create 20 000 new
jobs, the $31.9 billion increase in Canadian and American exports to Mexico will
translate into 638 000 new jobs.
† Section (f) of Article 102 of the agreement addresses this question in general terms.
According to this section, the objectives of the agreement include the establishment of 'a
framework for further trilateral, *regional* and multilateral cooperation to *expand and
enhance* the benefits of this Agreement.' More specifically, to accommodate its future
expansion, NAFTA includes an 'accession clause': '1. Any country or group of countries
may accede to this Agreement subject to such terms and conditions as may be agreed
between such country or countries and the Commission and following approval in
accordance with the applicable approval procedures of each country. 2. This Agreement
shall not apply as between any Party and any acceding country or group of countries if, at
the time of accession, either does not consent to such application' (Article 2205).
Inclusion of this clause is, by itself, an indication that NAFTA's expansion is a
possibility.
§ An indication of this was given in the official communiqué issued at the conclusion of a
visit by President Patricio Aylwin of Chile to Washington in May 1992. According to the
communiqué, the two countries will negotiate a free trade agreement after the completion
of the North American Free Trade Agreement. 'Free-Trade Pact Talks with Chile to
Await North American Pact', *The Wall Street Journal*, 14 May 1992, p. A2.

admitted jointly, in tandem with the above countries. The addition of Chile, Venezuela and any two or three Central American countries will add little to the size of NAFTA but much to its diversity and potential. Furthermore, and maybe more significantly, it will open the door to the formation of a truly 'hemispheric' trading bloc, or even an economically integrated Western hemisphere. This is, indeed, a long-term prospect and there are many obstacles to overcome. Nevertheless, in a world dominated by regionalism, the Western hemisphere countries might not have much of a choice but to pursue it.

NOTES AND REFERENCES

1. For an elaboration of Canadian opposition to the Canada–U.S. FTA and NAFTA see Philip Raworth, 'A North American Common Market: A Canadian Perspective', in Khosrow Fatemi (ed.), *International Trade and Finance: A North American Perspective* (New York: Praeger Publishers, 1988) pp. 270–84. It should be noted that this paper was written during the time that the negotiations for the Canada–U.S. FTA were in progress. The author's arguments are primarily socio-cultural (Canadian identity and Canada's close ties with Europe) and political (sovereignty). However, he does raise some economic concerns as well.
2. Article 13 of the original Single European Act. For further discussion of this issue, see David R. Cameron, 'The 1992 Initiative: Causes and Consequences', in Alberta M. Sbragia (ed.), *Euro-Politics: Institutions and Policymaking in the 'New' European Community* (Washington, DC: The Brookings Institution, 1992) pp. 23–74.
3. *North American Free Trade Agreement* (Washington, DC: U.S. Special Trade Representative's Office) Chapter One, Article 102.
4. Jim Giermanski, Kelly S. Kirkland, Eduardo Martinez, David M. Neipert and Tom Tetzel, *U.S. Trucking in Mexico: A Free-Trade Issue* (Laredo, TX: Laredo State University, September 1990) p. 31.

2 The Coming Debate on NAFTA

Sidney Weintraub

2.1 INTRODUCTION

The text of the proposed North American Free Trade Agreement (NAFTA) was submitted to the Congress by President Bush on 18 September 1992. Based on the timetable prescribed in the fast-track legislation under which the agreement was negotiated, the main congressional debate on NAFTA was thus reserved for the enabling legislation to be taken up in the post-electoral Congress. If approved by the three legislatures – Canada, Mexico and the United States – NAFTA will go into effect on 1 January 1994. This chapter examines the debate on NAFTA as it is shaping up in the United States.

The agreement is quite lengthy, more than 2000 pages. There are a number of reasons for this, some unexceptionable and others the consequence of the fact that the agreement is not 'clean', that is, it is not free trade without exceptions. The political process did not permit pure free trade in any of the three countries. The length of the agreement was also made necessary by its scope and because it contains detailed annexes on the winding down over time of restrictions of all three parties. The separate annex on the schedule for tariff reductions, while long, provides precision that should be welcome to traders. On the other hand, the rules of origin, especially for automobile products and textiles and apparel, are complex because they contain many protective elements limiting free trade to products containing elaborate constructs of what can be defined as North American content. The provisions in these two sectors also required the setting forth of changes from one protective system to another.

The agreement contains an annex with transitional commitments of the three parties to NAFTA with respect to three chapters, those dealing with investment, cross-border trade in services and financial services. These so-called reservations contain predominantly the change from Mexico's protected environment to what will be mostly an open market after a period of between 10 and 15 years. These are not restrictive provisions, but rather ones that set the pace of liberalization. They are, in fact, among the most gratifying provisions of NAFTA. The agreement is mostly liberalizing, although there are major exclusions and some backsliding. The backsliding is largely in the rules of origin under which free trade is limited in a variety of ways to production within the North American area.

The main Mexican exclusion is for trade and foreign investment in energy, for which Mexico insisted that the following clause be inserted in the agreement (Article 601(1)): 'The Parties confirm their full respect for their Constitutions'. Canada excluded cultural industries from the provisions of NAFTA, as it had done earlier in the Canada–U.S. FTA. The United States did not include coastal shipping in the earlier agreement with Canada or in NAFTA. There is a durable shipping lobby in the United States which manages to retain its protection under all circumstances. Of these three exclusions, the most significant is for Mexico's energy sector because this activity encompasses a significant proportion of Mexico's total investment and trade. The energy exclusion limits the benefits the United States might have wished to obtain from security of oil supply from a neighbor.

Various themes were consciously omitted from the agreement. Other than provisions for the temporary entry of business persons, the agreement contains nothing on immigration. There is nothing in the agreement on labor standards and workplace conditions. These are reserved for what has come to be known as the 'parallel track'. In contrast, environmental issues are included both in the agreement and in parallel-track understandings. Provisions on adjustment assistance – retraining and financial compensation of workers who lose their jobs because of NAFTA – are not included in the agreement, but this theme was always seen as appropriate for separate national legislation. The form and the funding for adjustment assistance will be a major issue in the debate on the enabling legislation.

Section 2.2 will provide more information on the structure and contents of the agreement. The provisions of the agreement as they deal with particular sectors will be touched on, but not set forth in elaborate detail. Other chapters in this volume deal with two key sectors – automotive, and textiles and apparel. The material on the contents of the agreement will be followed by a discussion of what is missing in the agreement. This is a necessary prelude to the arguments favoring or opposing congressional approval of NAFTA, which will follow in

Section 2.4.* My judgment is that NAFTA should be approved by the U.S. Congress. Both major presidential candidates agree with this conclusion. Certain features may need clarification in the enabling legislation, and this is within the prerogative of the U.S. Congress working with the president.

2.2 THE CONTENT AND STRUCTURE OF NAFTA

The proposed agreement has a preamble and 22 chapters.† The chapters, in turn, have their own annexes, either for clarification purposes or to list permanent exceptions from what is contained in the chapters. For example, Chapter 3 (national treatment and market access) has separate annexes on the automotive sector and textiles and apparel; and Chapter 7 (agriculture) has two subchapters, on market access and phytosanitary regulations. The exceptions in the annexes to the individual chapters are largely permanent. Mexico's declaration of nonapplicability of key undertakings in the energy chapter (chapter 6) is written into the annexes to this chapter – Mexico declared that the foreign investment provisions of NAFTA are not applicable to energy and refused to agree to the sharing of oil with the other two countries if there are production cutbacks.

There is a jerry-built quality to the structure. The intent clearly was to have a core agreement of provisions applicable to all three countries. The exceptions to these core understandings are contained in the chapter annexes. The transitory reservations are provided in a separate section. Finally, the tariff-reduction schedule has its own section. At the end of the day, after 10 to 15 years when the transition period runs its course, what will remain is the core, coupled with the permanent exceptions, while the reservations and the tariff schedules disappear. This treatment is not fully consistent throughout and one therefore gains the impression that there may have been some last-hour changes

---

* To be precise, Congress does not vote on NAFTA as such, but on the implementing legislation.
† The chapter headings are as follows: objectives, general definitions, national treatment and market access, rules of origin, customs procedures, energy, agriculture, emergency action, standards-related measures, government procurement, investment, cross-border trade in services, telecommunications, financial services, competition policy, temporary entry for business people, intellectual property, public notification and administration of laws, review and dispute settlement in antidumping and countervailing duty matters, institutional arrangements and dispute settlement procedures, exceptions, and final provisions.

of opinion about how to structure the formal text. By having a core and then separate material that is applicable only to the three countries, the negotiators hoped to simplify later accession negotiations.

The accession clause is quite general, leaving entry into the agreement open to other countries, without geographic limit, but making it clear that each of the three parties to the agreement must follow its own procedures for accepting new members. The expectation is that the first applicants would be Latin American and Caribbean countries. Indeed, Chile has already made known its desire to accede. However, by not limiting membership to countries in the Western Hemisphere, accession is nominally open to Asian and Pacific countries, such as Australia and New Zealand, both of which are in the midst of debates about their future trade relations with countries in their own region and with North America.

The agreement incorporates other agreements in a variety of ways. The provisions of the General Agreement on Tariffs and Trade (GATT) are incorporated throughout. The purpose, apart from simplifying the text, is to signal that NAFTA is consistent with the GATT framework. However, in the event of inconsistencies between GATT and NAFTA, the provisions of NAFTA prevail. NAFTA also incorporates a number of international environmental agreements – on trade in endangered species, the use of substances that deplete the ozone layer, and transboundary movement of hazardous wastes – and in these cases the environmental agreements prevail in the case of inconsistency with NAFTA. The motive was to mollify the environmentalists by assuring them that nothing in NAFTA is intended to despoil the environment. This is an important issue for congressional consideration. In addition, several features of the Canada–U.S. FTA are incorporated. The automotive provisions of the Canada–U.S. FTA, which in turn continued various Canadian safeguards from the 1965 Canada–U.S. automotive agreement, are largely retained. So, too, are the letters of commitment extracted by the Canadian government from the big-three automobile producers at the time of the 1965 Canada–U.S. automotive agreement. Consequently, following the transition period, the Mexican automotive market is apt to be more open to imports from the United States than is the Canadian market.

While the agreement makes its obeisances to vested interests in the three countries – some of which will be noted later – its overall thrust is liberalizing. There will be free trade in most goods and many services after 10 years. For some products, those which are sensitive for a variety of reasons, the transition to free trade will be 15 years. For example, the transition period is 15 years for corn and dry bean imports into Mexico and orange juice and sugar imports into the United States. In the case of used automobiles, the Mexican market will open gradually but not be fully free until 1 January 2019, a full 25 years after NAFTA is scheduled to go into effect.

The liberalization in some areas is indeed remarkable, certainly viewed in the light of previous history in Mexico and the United States, and their earlier economic relationship, or based on what has been accomplished in other trade negotiations. As a general rule, there will be free trade in agriculture between Mexico and the United States after 10 years, or running up to 15 years for a few sensitive products. This point merits emphasis. In other regional integration arrangements, trade in agriculture was either completely or mostly omitted (as in the European Free Trade Association or even in the Canada–U.S. FTA) or dealt with in a highly protective manner (as under the common agricultural policy of the European Community). GATT itself has barely been able to come to grips with liberalizing agricultural trade.

Yet, despite the great sensitivity of this sector, Mexico and the United States reached what can only be called breathtaking breakthroughs. The agreement will compel Mexico to make the most of the 15-year transition for opening the corn market by providing alternative opportunities for what may be as many as three million persons who now eke out a subsistence existence in rainfed corn-growing areas. Mexico undoubtedly had concluded that this adjustment was necessary in any event, but the courage to move ahead is still remarkable.

Mexico's investment regime had long been restrictive, although there has been much liberalization during the past decade based primarily on regulatory changes. The basic law itself, which was enacted in 1973, has a remarkably revealing title: the law to *promote* Mexican investment and *regulate* foreign investment (my emphasis). Most of the restrictive features will disappear. Mexico will retain the right to screen acquisitions that are higher than a specified threshold ($25 million at first, phased up to $150 million after 10 years); in this respect, Mexico mimicked the treatment that Canada received in the Canada–U.S. FTA. In addition, Mexico will gradually phase out its performance requirements, namely, those conditions which require foreign investors to meet specified export levels, reach minimum domestic content in production, or prefer domestic sources in their procurement. The investment understandings require Mexico to change its laws as they now affect the other two parties to the agreement.

Mexico, until now, has been largely closed to foreign banks.* This changes under the agreement. Foreign share limits will be progressively increased. There are similar provisions for U.S. and Canadian insurance firms and companies dealing in securities in Mexico. Cross-border trade in other services will similarly be opened. The basic principle here is national treatment, that is, treatment no less favorable than treatment of a country's nationals. There are exceptions to this, particularly relating to local levels of government, but the provisions are more comprehensive than any liberalization achieved thus far in GATT negotiations or even in the Canada–U.S. FTA.

Cross-border land transportation, such as bussing and trucking services, have long been restricted between Mexico and the United States. Cargo moving by truck between Mexico and the United States typically had to be transferred to the rig of the other country at the border, or in the area immediately adjacent to the border. Back hauls were generally nonexistent. This will change gradually over a transition period. Three years after signature of the agreement, that is, in December 1995 if the agreement is signed in December 1992, trucks from the three member countries will be permitted to make cross-border deliveries and pick up cargo in U.S. and Mexican border states. Then, six years after the agreement goes into effect, that is, 1 January 2000 under the current schedule, each country will open its territory to trucks of the other two. Bus companies will be permitted to provide service to any part of other member countries three years after the agreement goes into effect. The proportion of permissible U.S. and Canadian ownership of bus and truck companies in Mexico will increase gradually to 100 per cent after 10 years.

This seemingly technical provision will sharply open competition in land transportation and should lower the cost of moving cargo. There will be winners and losers among bus and truck companies, as one would expect from increased competition, but the cost of moving merchandise should go down. The agreement will require making technical and safety standards of the three countries compatible for land transport. The Mexican constitution provides for national ownership of the rail system and this will not change under the agreement. The land aspects of marine transport will be opened to foreign investment.

---

* The word 'largely' was used because Citibank has long operated in Mexico and was not expropriated in 1982 when other Mexican banks were taken over.

The Canada–U.S. FTA broke new ground in establishing a number of procedures for the settlement of disputes. These are carried over into NAFTA. The most innovative of the various dispute-settlement procedures is that in Chapter 19, which provides a mechanism for binational panels to review final antidumping and countervailing duty decisions. These reviews, based on a complaint of one of the parties, are designed to substitute for domestic judicial review. Five-member panels chosen from a roster of experts will determine whether the country imposing the antidumping or countervailing duty complied with its own laws and regulations. As in the Canada–U.S. FTA, panel decisions are binding, subject only to an extraordinary challenge to a panel of three judges.

The use of contingent protection inherent in petitions alleging dumping and subsidies has mushroomed in the United States in recent years. In 1990, more than 90 per cent of the number of trade remedy petitions submitted against imports into the United States were for antidumping or countervailing duties, mostly antidumping duties (Coughlin, 1991, p. 11).* One of the Canada's main objectives in seeking free trade with the United States was to find a way to deal with this protection. The use of binding arbitration to substitute for judicial review was intended to be a temporary arrangement until more permanent procedures could be devised over a five- to seven-year period. As it worked out, Canada has won a number of cases in which the panels found that U.S. administrative decisions which imposed countervailing and antidumping duties were improperly reached under U.S. law. In the one case taken to an extraordinary challenge, dealing with Canadian pork exports to the United States, the judicial panel unanimously upheld the conclusion of the earlier panel that favored Canada. The binding arbitration procedure has worked more effectively than was anticipated when it was announced.

The U.S. negotiators at first resisted extending this procedure to Mexico, ostensibly because of the lack of clear due process in Mexican handling of subsidy/dumping cases. There may also have been some concern that Mexican appeals to arbitration panels would be as successful as those made by Canada. In any event, the procedure is part of the NAFTA agreement and it does require Mexico to change

---

* The kinds of petitions examined in Coughlin were escape clause cases and petitions under section 301 of U.S. trade legislation, in addition to antidumping and countervaling duty requests. The 90 per cent proportion for antidumping and countervailing duty cases has been typical for most, although not all, years since 1980.

many of its practices so that there will be a paper trail to determine whether Mexico complied with its own laws and regulations when it imposed antidumping or countervailing duties. These few provisions of NAFTA were described to illustrate the liberalizing thrust of the agreement. Other sections could be cited in support of this conclusion, such as those dealing with intellectual property, government procurement, greater transparency in setting technical standards, and access to telecommunications services. The agreement can be criticized for its omissions and backsliding. Many vested interests will object to its liberalizing features which open all three countries to greater competition in the provision of goods and many services. It cannot be criticized legitimately for being timid in seeking to integrate the North American market.

Having said this, the main protectionist aspect of the agreement is in the rules of origin. A free trade agreement, in contrast with a customs union, has no common external tariff for all its members. Because each member country retains its own tariff against non-members, rules are needed to prevent a third-party export to a low-tariff member country which is later transshipped to the high tariff country. Free trade is intended to apply only to goods originating in the member countries. If the goods do not originate in the member countries, the basic rule of origin to determine whether a good is eligible for free trade is a substantial transformation test, that is, whether the basic nature of a product was transformed in a member country before exportation to another member. If there is a change in tariff classification, this is generally considered to have required a substantial transformation, making the good eligible for free trade.

This rule, while not always clear-cut, is relatively simple. However, rules of origin are also used as protective devices. In these cases, they become similar to domestic-content provisions – only in this case, regional NAFTA content. The U.S. negotiators insisted that Mexico remove national-content provisions in its treatment of various products, particularly in the automotive industry, but then insisted on stringent regional-content provisions in this same industry. It is fair to state that without a high regional content in this sector, there could not have been an agreement. The U.S. concern was that Japan would establish simple assembly plants in Mexico as a platform for export to the United States. The rules of origin are also quite strict for trade in textiles and apparel.

Both these industries are treated in separate chapters in this volume. Simplifying here for the sake of brevity, the regional content to make a vehicle eligible for free trade under NAFTA will rise gradually to 62.5 per cent after the year 2002 for passenger cars and 60 per cent for other vehicles and parts. This is calculated under what is defined in the agreement as the net-cost formula for measuring regional value. The regional value for vehicles to be eligible for free trade in the Canada–U.S. FTA is 50 per cent. However, the method of calculation is different in the FTA and in NAFTA, so the two percentages are not directly comparable. For one thing, the rollup technique of the FTA counted regional value of inputs as zero or 100 per cent depending on whether the calculation was below or above 50 per cent. NAFTA calls for tracing the origin of inputs to get precise values. This may prove to be onerous for a passenger car that carries a most-favored-nation (MFN) tariff of only 2.5 per cent for entry into the United States.

It is difficult to assess the impact of NAFTA on the automobile trade in North America. The principal competition in this industry in the United States and Canada has come not from Mexico, but from imports from Japan and Japanese transplants in the United States and, to a lesser extent, from European production. However, Mexican exports of finished cars and parts have been rising and together now constitute the most important export to the United States, by value, after crude oil.

The United Auto Workers union vehemently opposes NAFTA because of concern that it will lead to a shift of parts and vehicle production to Mexico to take advantage of low wages. Even now, without free trade, the biggest growth in production and employment in the maquiladora industry has been in the transport sector, largely for the manufacture of parts for use in vehicles assembled in the United States. One prominent U.S. analyst expects North American production and sales integration to be a net plus for the U.S. industry in competition with Japan because of the growth in the Mexican market itself plus a division of labor under which, less-expensive, entry-level cars and trucks will be produced in Mexico, using parts from the maquiladora, and more expensive vehicles will be produced in the United States and Canada (Womack, Jones and Roos, 1991).

The textile and apparel industries are equally, if not more, sensitive than the automotive industry because of their sheer amount of employment and their geographic dispersion across the United States. The two elements of the industry – the mill and the apparel sectors – together had 29 000 plants spread throughout every state in the United States and they employed almost 1.4 million people in 1990 (Hufbauer and Schott, 1992, pp. 264–6). The rule of origin in these two sectors has been referred to variously as 'yarn forward' or 'triple transformation'. To be eligible for free trade, yarns must be made from material produced in North America, the fabrics must be made from eligible yarns and the products from qualified fabrics. There are exceptions for fabrics not produced in North America, such as silk. There are also exceptions for tariff rate quotas, that is preferential treatment up to a certain amount, for yarns, fabrics and products that do not meet the rules of origin.

This sector is complicated by the existence of the multifiber arrangement (MFA), an arrangement accepted by the GATT Contracting Parties, under which textile and fabric trade is now subject to import quotas in most industrial countries. One outcome of the Uruguay Round, if it is concluded, could be a gradual phasing out of the MFA. If this occurs, the provisions of NAFTA will have a different outcome for the three member countries than if they alone enjoy free trade in North America. Under NAFTA, tariffs will be phased out over 10 years for products that meet the rules of origin. The United States will immediately remove its quotas against Mexico on textiles and apparel that meet the rules of origin (Mexico does not have import quotas under the MFA) and will gradually phase out these quotas for Mexican products that do not meet the rules of origin. The expectation, if the MFA is retained, is that U.S. exports of yarns and fabrics to Mexico will increase, but that Mexican exports of finished products will increase as investment grows to take advantage of the low wages in this labor-intensive activity (Hufbauer and Schott, 1992, p. 278).

This discussion of the liberalizing and restrictive aspects of NAFTA as negotiated merely scratches the surface of what is, after all, a complex document in which provisions from any one chapter must be read in conjunction with related material in other chapters. An analyst can focus on the restrictive elements of the agreement and conclude that it is less liberalizing than would be the ideal under Article xxiv of GATT, which sets the criteria for internationally acceptable free trade agreements and customs union. Or, the stress can be placed on the liberalizing features of the agreement and accept that exceptions were required because of internal political, economic and social pressures in each of the three countries. My own reading is that NAFTA is considerably more liberalizing than it is restrictive; and that it goes much further in this direction than unbiased observers would have predicted as little as six months ago.

Beyond this, it would be an oversimplification to take NAFTA literally by its title as a free 'trade' agreement. It is that, mostly, but it deals with much more than trade. Mexico expects NAFTA to stimulate much investment, both foreign and domestic, to take advantage of the enlarged, mostly barrier-free market in North America. Mexico is already altering its investment laws to make them consistent with what is agreed in NAFTA. Similarly, Mexico is altering its statutes and procedures to make its determination of countervailing and antidumping duty cases more transparent. If U.S. laws on branch banking and the separation of commercial and investment banking change, Mexican and Canadian bankers will receive national treatment. U.S. financial institutions, now mostly barred from Mexico, should begin to operate there. Mexico has already strengthened its laws on the protection of industrial and intellectual property.

NAFTA, in other words, for better or worse, is apt to lead over time to considerable change in the way the three countries in North America, and their nationals, do business and interact with each other. The initial impact will be greatest on Mexico because it has the smallest of the three economies, but the effects on the other two will increase over time. This is particularly the case if the Mexican economy grows rapidly because U.S. exports to and investment there are determined primarily by the level of income in Mexico. Whatever economic growth and social transformation Mexico accomplishes will be the result primarily of its own policies, but NAFTA can play a reinforcing role.

2.3 WHAT IS MISSING

When Mexico took the initiative for a free trade agreement with the United States, both Mexican officials and private citizens suggested that migration issues should be included in the negotiations. The reasoning was twofold: this is an important issue in the relationship and some formal understandings would be desirable; and if capital movements were to be included, as of course they were, then the movement of the other major factor of production – labor – should also be part of the agreement. The Mexican authorities never made clear what they had in mind: a temporary worker agreement; movement of Mexican labor to provide services, such as when a Mexican entrepreneur was awarded a construction contract in the United States; or perhaps some assurances on the protection of the human rights of Mexicans in the United States, whether there legally or without documents.

This was not accepted by the United States.* Yet, while not explicitly included, many Americans support NAFTA because they are convinced that economic development is the only sure way to slow down undocumented immigration from Mexico. Is this a vain hope? It probably is, for the short term; but it may not be for the long term (see Commission for the Study of International Migration and Cooperative Economic Development, 1990). It is hard to define the long term. It could take more than 100 years for per capita incomes in the two countries to equalize, but this may not be necessary to reduce the incentive to emigrate. If income and employment in Mexico were on a steady upward path, coupled with more equal distribution of the benefits than in the past, this sense of hope for the future may be sufficient to stanch emigration pressure. Theories of international migration are a separate theme not suitable to this chapter. And, in any event, the theme was not taken up in the negotiations.

During the debate on the extension of fast-track procedures for the Mexican negotiation, the administration made a number of promises to Congress on themes that would be included in the negotiations, either directly or in what was referred to as a 'parallel track' (see Bush, 1991b). The president made three types of commitments, achieving 'a balance that recognizes the need to preserve the environment, protect worker safety, and facilitate adjustment'.

Only the first of these, on environmental issues, was included in the agreement. In addition, environmental understandings were negotiated on the parallel track. The second, dealing with workers' safety, is not included in the agreement; it was negotiated solely on the parallel track. The third, on obtaining assured financing for workers' adjustment to dislocations resulting from free trade, is a subject to be worked out in the enabling legislation when it is taken up in 1993. The Bush administration, when it forwarded the NAFTA agreement to Congress in September 1992, also sent forward a report documenting its actions in these three areas (Office of the U.S. Trade Representative, 1992).

Full discussion of these three issues will be deferred until Section 2.4 because they will figure prominently in the congressional decision to approve the enabling legislation required to put NAFTA into effect. What can appropriately be noted here is that there is disagreement in

---

* Indeed, I know from personal discussions that the Mexican authorities were told there would be no free trade discussions if they insisted on migration provisions, other than those related to temporary entry for business people.

the United States about whether there is sufficient assurance that Mexico can or will enforce its environmental undertakings, either those in the agreement itself or others negotiated on the parallel track; and whether a parallel track understanding on Mexican enforcement of worker rights is sufficient, that is, whether there should be something on this in the NAFTA itself or in a side agreement.

One additional theme arose during the debate on the extension of fast track, namely, whether it was appropriate to negotiate with a country that has a poor record of enforcement of human rights, including electoral rights. This theme was never appropriate for inclusion in the agreement itself, or even in parallel understandings. However, there is an anomaly here. One Mexican motive for wanting something on migration in the agreement, or on a parallel track, was to assure protection of the rights of its citizens in the United States. This was dismissed by the U.S. government.*

There was a sense in Mexico, evident in the Mexican press, that U.S. opponents of NAFTA were deliberately overloading the agreement in order to frustrate its conclusion. Thus, inclusion of environmental provisions was first resisted by both governments. This resistance collapsed when it was clear that environmental provisions were essential if there was to be an agreement. The same is true of workers' rights; this too was resisted by both governments and never made it into the agreement. While human rights and electoral issues were never seriously considered for inclusion in the agreement, it is worthy of note that President Carlos Salinas de Gortari has established commissions to deal with both these issues inside Mexico. At least one congressman argued for inclusion of drug enforcement issues in the negotiations. They were not included, but cooperation in this area has been satisfactory in recent years.

Many of the concerns raised both in Mexico and the United States during the fast-track debate have thus been addressed, although not necessarily to the satisfaction of critics. One of two conclusions can be reached from this history – either that governments will do only what is necessary unto the moment, or that intense interaction between Mexico and the United States does provide the best framework for progress on sensitive issues. I suspect that there is some truth in both these conclusions.

---

* My view is that Mexican democracy, particularly effective suffrage, is likely to be fostered by NAFTA.

2.4 THE CASES FOR AND AGAINST NAFTA

The main argument against NAFTA is that low Mexican wages, about one-eighth, on average, of those in the United States, will attract much investment and lead to the loss of U.S. jobs. The argument is carried one step further in that even if there is a net gain in U.S. jobs over time, there will be no effective compensation, no meaningful retraining or financial assistance for those U.S. workers and families who are disadvantaged. The likely losers are low-wage workers in relatively labor-intensive industries. The best example is the apparel industry.

The assertion that Mexico does not adequately enforce either its labor or environmental laws fits partially into this low-wage argument. The reasoning is that this lax enforcement gives producers in Mexico a further cost advantage in competition with U.S. industry. The word partially is used in the previous sentence because environmentalists and supporters of workers' rights believe in their positions apart from the competitive issue.

Arguments against NAFTA are often nuanced. For example, the Office of Technology Assessment issued what is at best a conflicting report which states that failure to approve NAFTA could actually increase the pressures on less-skilled U.S. workers, but that approval of NAFTA without significant domestic reforms in the United States 'might be tantamount to ratifying the mismanagement of economic integration' (U.S. Congress, 1992, p. 9).*

The net job-loss argument is at best ambiguous. Analysis of the effects of NAFTA gave much life to the technique of computable general equilibrium (CGE) modeling. There are literally dozens of CGE models looking at NAFTA from the vantage point of each of the three countries and of North America as a whole. They will not all be cited here, but many are contained in the 1992 U.S. International Trade Commission publication cited among the references. Lustig, Bosworth, and Lawrence (1992) contains an analysis of many CGE and other studies. The general conclusion – indeed, almost without exception among CGE studies – is that the United States will, on an economy-wide basis, gain jobs from NAFTA. The figures are modest in the short term, perhaps in the tens of thousands to hundreds of

---

* This analysis has all the earmarks of overloading NAFTA by indicating that it should be approved only if a wide panoply of domestic policy reforms are instituted first, or at least simultaneously. Many of these reforms are undoubtedly needed, but why put the entire burden on the NAFTA approval process?

thousands – numbers that fall within the margin of error for an economy the size of the United States – but even these figures are deceptive. This is because the conclusions are based on assumptions about returns to scale in the U.S. industries that benefit, what dynamic considerations are taken into account, and how one models the job-effects of increased competition. If different assumptions are made, the numbers would change.

Many of the models concern themselves primarily with examining potential intersectoral shifts rather than overall effects on the grounds that U.S. macroeconomic policy is the principal determinant of U.S. employment. The Hufbauer–Schott calculations, which are not based on a CGE model, project the *net* creation of 130 000 additional jobs from NAFTA, based on job creation from increased U.S. exports (p. 55). However, Hufbauer–Schott also estimate that the *gross* number of displaced workers from increased imports could reach 112 000 (p. 60). The Economic Policy Institute, a labor-related think tank, states that estimates of net job changes from various studies range between a gain of 130 000 to a loss of 900 000 (Faux and Lee, 1992, p. 5).* All these estimates, particularly as they pertain to the short run, should be approached with skepticism.

One job-related argument against NAFTA is that it would reinforce efforts by U.S. employers to use a low-wage strategy, that is, to use Mexican production facilities to remain competitive rather than adopt a high-wage, high-productivity strategy (U.S. Congress, 1992, p. 7). This turns the argument of NAFTA supporters on its head. For the supporters, the argument is that defeat of NAFTA would safeguard precisely the low-wage industries in which the United States cannot compete without substantial import protection, such as apparel manufacture. The higher-wage, higher-productivity industries do not need this protection. Japan also exports the low-skill, low-wage aspects of manufacturing in coproduction arrangements with a number of Asian countries. This is precisely the logic of regional integration, to take advantage of differences in factor endowments as the basis for competitive production.

Once again, the opponents of NAFTA seek to put all the burden of domestic U.S. policy on NAFTA. Imports from Mexico represent about one-half of one per cent of U.S. GNP and policy must be made

---

* The high estimates are not based on CGE modeling. I find this range troubling. It is a one-assertion, one-vote statement.

in the other 99.5 per cent. It is not really the management of NAFTA that is at issue, as asserted by the Office of Technology Assessment, but the management of the U.S. economy – NAFTA or no NAFTA. The case in favor of NAFTA is that the Mexican economy is already highly integrated with the U.S. economy, much as is the Canadian economy. This shows up in the importance of the two North American countries as markets for U.S. products: U.S. merchandise exports to the two countries in 1991 were Canada $85 billion and Mexico $33 billion, the first and third U.S. markets (Japan was second at $48 billion). U.S. shipments to Canada and Mexico together constituted 28 per cent of all U.S. merchandise exports in 1991. U.S. investment is also high in these two countries. Much of the trade in manufactured goods – indeed, more than half – is between related companies, and much of this is in intermediate goods. A free trade agreement is hardly needed merely to remove tariffs because they are already low: they average about 3 per cent for U.S. imports from the two countries. NAFTA, viewed in efficiency terms, is designed to augment the economic integration that is already taking place by providing some certainty of treatment for investment, coproduction and integrated sales in North America.

Concern that the U.S. market will be swamped by cheap imports from Mexico has little basis in the evidence to date. What has happened instead is that U.S. exports to Mexico have increased much more rapidly than U.S. imports from Mexico once the Mexican economy began to recover. This can be seen in Table 2.1. There is also evidence from econometric analysis that Mexico's marginal propensity to import from the United States is greater than the U.S. marginal propensity to import from Mexico (Cabeza, 1991). This suggests that the U.S. trade surplus with Mexico will continue to grow as the Mexican economy grows. The United States thus has a major trade and economic stake in Mexico's economic health.

The evidence, I believe, is that the proponents of NAFTA have stronger economic arguments than the opponents. This is certainly true for Mexico, but also for the U.S. economy.

The weight of the evidence on the social issues, however, is less decisive. It is on perceptions of how well the agreement itself and parallel understandings deal with these issues, and on how effectively the U.S. government can handle matters of worker dislocation – which will occur regardless of whether there is *net* job creation – on which approval of NAFTA will hinge. This is the message from the leadership of the Democratic Party and from Bill Clinton himself (Clinton, 1992).

Table 2.1   U.S. merchandise trade with Mexico, 1985–91 (billions of dollars)

	U.S. exports	U.S. imports	Balance
1985	13.6	19.1	−5.5
1986	12.4	17.3	−4.9
1987	14.6	20.3	−5.7
1988	20.6	23.2	−2.6
1989	25.0	27.2	−2.2
1990	28.3	30.2	−1.9
1991	33.3	31.2	2.1

Source: U.S. Department of Commerce, *U.S. Foreign Trade Highlights, 1991.*

The U.S. administration has enumerated the following ways in which the NAFTA agreement promotes environmental protection (Office of the U.S. Trade Representative, 1992, table 7, pp. 7–8):

- NAFTA maintains federal and state environmental standards and the U.S. right to ban nonconforming imports;
- it allows the parties, including states and cities, to enact standards that are more strict than international and national standards;
- it allows each party to choose the level of protection of human, animal, or plant life it considers appropriate;
- it encourages upward harmonization of standards;
- it allows parties to adopt regulations and sanitary and phytosanitary standards based on available information when there is insufficient evidence to conduct a risk assessment;
- it allows each country and its nationals to review and comment on proposed regulatory actions;
- it establishes a committee on standards-related measures to facilitate compatibility of standards; and
- it establishes a committee on sanitary and phytosanitary measures to enhance food safety.

In addition, when a dispute arises that raises factual questions about a country's standards, or measures taken under international environmental agreements, the complaining party may submit the issue for resolution under the NAFTA dispute-settlement procedures, where the complaining party has the burden of proof. Because workers' rights issues are not part of NAFTA itself, disputes over this issue cannot be submitted to the dispute-settlement provisions of NAFTA.

Finally, the two governments, using the parallel track, released an integrated border environmental plan on 25 February 1992.

The argument of those who believe that the environmental provisions of NAFTA do not go far enough is based on their conviction that while formal environmental standards in Mexico are generally comparable to those in the United States, they are often ignored. Despite increased attention to and budgets for environmental issues, there is also concern that the Mexican authorities lack the wherewithal to bring about stricter enforcement. One proposal that has been made to deal with this perceived shortcoming is to establish a binational commission, with adequate resources and substantial power to impose penalties on violators, to prevent environmental degradation (see U.S. Congress, 1992, p. 8; Clinton, 1992). President Clinton has also suggested negotiation of a parallel agreement among the three countries to permit nationals to bring suit in their own courts when they believe environmental protection laws and regulations are not being enforced.

There is no consensus among environmental groups in the United States about the desirability of NAFTA. Some organizations support NAFTA, while others are withholding support until corrections of the type noted above are made. NAFTA supporters among environmentalists argue that the progress that has been made is substantial; and, more importantly, that the border plan and the inclusion of environmental issues in a major trade agreement is too important to discard. The other side of the argument is that if there are weaknesses in the enforcement of environmental laws by the Mexican authorities, this is the moment to take corrective measures.

There is a long history of attempts to include labor rights and workplace health and safety regulations in international trade agreements. These efforts have not gone very far because of differences among nations in their levels of development and concern by developing nations that the introduction of this issue by developed countries is largely a subterfuge for eliminating competition. The United States has taken unilateral action in this field in the legislation that provides for preferential treatment of designated imports from developing countries. Basic workers' rights in this legislation are defined as including the right of association; the right to organize and bargain collectively; a prohibition on forced labor; a minimum age for employment of children; and acceptable conditions of employment with respect to minimum wages, hours of work and occupational safety and health (Weintraub and Gilbreath, 1992, p. 48).

Mexico and the United States, using the parallel track, first concluded a memorandum of understanding on labor cooperation, and then in September 1992 concluded an agreement to establish a consultative commission on labor matters (Office of the U.S. Trade Representative, 1992). The purpose of the agreement is to exchange information on workers' rights and labor standards. The agreement explicitly prohibits either party from undertaking actions in the other that are exclusively entrusted to national authorities.

The response of those who believe these steps are inadequate is that more must be done to assure Mexican enforcement of its own labor laws. There is a parallelism here with environmental issues. The call for a binational commission on workers' rights contained in the study of the Office of Technology Assessment (p. 8) has been met, although the powers of the commission are consultative. President Clinton has suggested, as with violations of environmental regulations, that nationals of the three countries be empowered to bring suit in their own courts when workplace standards are not being met. There may be an effort to bring workers' rights issues into the agreement in order to make them subject to the dispute-settlement provisions of NAFTA.

The third social issue that has arisen in the debate deals with the proposed criteria for retraining and other adjustment assistance for workers who are hurt as the result of NAFTA, either because of movement of plants or increased imports. Quite late in the day, in August 1992, President Bush made a proposal for $10 billion of funding over five years for what he called advancing skills through education and training. The program would replace both the Economic Dislocation and Worker Adjustment Assistance Act (EDWAA) and the Trade Adjustment Assistance Act (TAA). Of the total of $2 billion a year, President Bush proposed that a minimum of $335 million be allocated exclusively for NAFTA-related dislocations. This could be doubled up to $670 million if the Secretary of Labor ruled that more was needed to deal with NAFTA dislocations. The proposal said nothing about where the appropriations would come from.

It will clearly be necessary to deal with dislocations stemming from NAFTA in the implementing legislation. Whether this should be confined to NAFTA-related dislocations, trade dislocations general-ly, or retraining needs required by overall adjustments in the U.S. economy, will have to be sorted out. The amount of funding needed even for NAFTA-related dislocations alone is by no means a settled issue. Hufbauer and Schott, based on their estimate that up to 112 000 persons might be dislocated over several years as a result of NAFTA, conclude that a one-time appropriation of $900 million ($8000 each for 112 000 persons) might be sufficient (Hufbauer and Schott, 1992, p. 60). This is less than the amount proposed by President Bush. Where this funding would come from at a time of tight budgets and strong anti-tax sentiment is far from settled. Labor unions are almost certain to seek adjustment benefits as an entitlement on the ground that NAFTA-related problems were created by government policy and should be eased by assured government funding. This may turn out to be the most difficult issue in obtaining approval for NAFTA. There is disagreement among experts about the adequacy of the environmental provisions of NAFTA and the additional understandings reached in parallel track negotiations. As we have learned after years of international negotiation in GATT and the International Labor Organization, it is hard to come to grips with the enforcement of labor standards in countries other than one's own. The parallel-track negotiations on this issue have at least served to set standards based on what are generally acceptable national laws; a mechanism has been established to collect information about violations of these standards and a procedure now exists for consultations about these.

What has been accomplished in these two fields may be acceptable with only modest modifications. There is no comparable degree of agreement on which to build for adjustment assistance. Legislators in the United States can berate Mexico when it comes to the enforcement of its laws on the environment and labor rights. In the case of adjustment assistance, the criteria and the funding mostly come from within. This, I submit, is much harder for a politician.

2.5 CONCLUSIONS

The initiative for NAFTA came from Mexico only after profound changes took place in economic policy. Once development philosophy shifted from looking inward for the source of economic development to making exports of manufactures a key engine of growth, Mexico found it necessary to assure access to its main market in the United States. NAFTA, while not inevitable, was a logical step. It was also a way of obtaining reciprocity for the unilateral opening of the Mexican market that occurred during the 1980s.

For the United States, it would have been remarkable if the Mexican proposal for regional integration had been rejected. This proposal was coupled with a shift in political attitude, from a deliberate distancing of Mexico from the United States to a partnership. Many in the U.S. Congress were not enamored with the idea of negotiating free trade with Mexico, but approval of the negotiation under fast track was granted.

The moment of truth has now come. A proposed agreement has been submitted for approval. There is no dearth of analyses and these come to every conceivable conclusion. What cannot be denied, however, is that the proposed agreement is comprehensive. It is, on the whole, liberalizing. This has itself generated opposition from industries and workers who fear more open markets. The rules of origin in the two sectors discussed here, automobiles and textiles and apparel, reflect concessions to these national interests in the United States who are not prepared for full free trade. The agreement does not promise completely free trade even after the transitional period, but it clearly approaches this objective.

All of these matters will enter into the debate on the implementing legislation – from those who want more protection and those who want less, and from specific commodity interests on both sides of this argument. Yet, three issues have risen to the forefront of the debate. These are whether the agreement and its parallel negotiations go far enough in protecting the environment, whether the parallel-track understandings on workers' rights issues are adequate, and how best to deal with worker dislocation resulting from NAFTA even if, on the whole, the agreement helps to create jobs in the United States.

Just as it would have been remarkable had the U.S. administration spurned the Mexican initiative to negotiate a free trade agreement, so will it be extraordinary if Congress rejects the agreement.

Part II
Macroeconomic Issues

3 Introduction

In this part, three experts examine the macroeconomic issues of the North American Free Trade Agreement. In Chapter 4, Robert Shelburne compares NAFTA with the Southern expansion of the European Community in the 1980s when Greece, Spain and Portugal were admitted to the European Community. His comparison includes, *inter alia*, population, income and trade volume. Shelburne's contribution is particularly important because of the prevailing misconception among the general public, including some observers of North American issues, that NAFTA is analogous to the European Community. Shelburne argues that despite some similarities between the two cases, 'the economic integration of the EC is much more extensive than what is being planned for the NAFTA . . . [and therefore] one must be careful in making analogies between the two cases'. Nevertheless, he concludes that, if the experience of the EC is repeated in North America, 'the Spanish case seems most analogous to the Mexican case . . . [and] if the Spanish experience is duplicated, we would expect Mexican growth to increase, an increased trade deficit for Mexico matched by capital inflows from the U.S., significant increases in the volume of bilateral trade with much of it being intra-industry, some trade diversion and a very small convergence in income and productivity'.

In Chapter 5, M. Reza Vaghefi discusses NAFTA's potential as a base for a North American common market. His analysis includes a country-by-country discussion of economic and political issues which the member countries of NAFTA must face before the FTA's expansion into a common market can be seriously debated. Among the issues Vaghefi lists are nationalism and the fear in both Canada and Mexico of being 'swallowed whole, culturally and economically', a general mistrust of the United States by the other countries, and the real problems of an integration of unequal partners with the resultant 'asymmetric [inter]dependence' among the three economies.

Finally in this part, Chapter 6 is an examination of the trade picture in North America in the 'post-debt-crisis era' by Barry W. Poulson and Mohan N. Penubarti. The authors use historical expectations to set the framework for their analysis and then use time-series analysis to reach their conclusions. Poulson and Penubarti reach conflicting conclusions

41

with regard to the impact of the debt crisis of the early 1980s on intra-regional trade. On the one hand, they found 'no evidence of a significant change' in the patterns of trade between the United States and Mexico. Their study of the 'trade between Canada and Mexico, on the other hand, shows greater discontinuity in the period following the debt crises'.

4 The North American Free Trade Agreement: Comparisons with Southern EC Enlargement

Robert C. Shelburne*

4.1 INTRODUCTION

There is considerable interest in ascertaining the economic consequences of creating a North American Free Trade Agreement (NAFTA) by including Mexico in the U.S.–Canada FTA which was created in 1989. The NAFTA would be unique in that never before have nations with such divergent levels of per capita income agreed to eliminate their trade barriers. The economic rationale for such an agreement derives from the standard conclusions of international trade theory which show that free trade promotes economic efficiency and growth. In the United States, however, there is concern that such an agreement would further promote intersectoral specialization based on factor intensities and this would produce, by way of Stopler–Samuelson effects, falling real wages for unskilled labor. This is a particularly sensitive issue in the United States since the absolute real income of unskilled workers has been declining since the early 1980s. Numerous studies, such as those by Abowd and Freeman (1991), Borjas, Freeman and Katz (1991) and Wood (1991), attribute a significant proportion of this decline to increased trade. In addition, a NAFTA may create transitional adjustment pressures in many low-wage labor-intensive industries. Thus the primary groups in the United States that are likely to be harmed by NAFTA are the same groups

---

* The views expressed in this chapter are those of the author and may not necessarily reflect the positions or opinions of his employer, the U.S. Department of Labor.

43

that have already experienced a significant reduction in their real incomes and are now near the bottom of the economic ladder.

Numerous modeling efforts using a wide range of methodologies have been undertaken in order to estimate the economic consequences of NAFTA. Schoepfle and Perez-Lopez (1992) and Hinojosa-Ojeda and Robinson (1992) provide excellent surveys of these attempts. Although there is a general consensus among the traditional neoclassical (computable general equilibrium) trade models about the comparative static effects of the trade liberalizations inherent in NAFTA, there is very little agreement among the more ambitious models which attempt to model factor movements, induced technological change and macroeconomic dynamics. These modeling efforts are limited in that they attempt to extrapolate significantly beyond the observed data and must incorporate some qualitative changes for which there are no precedents. What is ideally needed is a similar historical experience that can be studied in order to provide researchers with a general guide as to what is likely to happen when countries of significantly different economic levels attempt to integrate economically.

The analogy of Mexico's incorporation into NAFTA and the southern enlargement of the European Community (EC) during the 1980s has been made numerous times both by government officials such as U.S. Trade Representative Carla Hills and academic researchers such as Jagdish Bhagwati and Rudiger Dornbusch. This paper addresses the appropriateness of this analogy and whether EC enlargement can provide any useful insights into a possible NAFTA. In Section 4.2 a comparison is made of the basic economic attributes of NAFTA and the EC, Section 4.3 discusses the institutional differences between the two cases, Section 4.4 reviews the results of EC enlargement, Section 4.5 attempts to relate the EC's experience to Mexico, and Section 4.6 provides a summary.

4.2 COMPARISON OF THE ECONOMIC ATTRIBUTES OF THE BLOCS

4.2.1 Population and Income Comparisons

The enlargement of the EC (Greece in 1981, Spain and Portugal in 1986) resulted in a 22 per cent increase in the population of the EC; the addition of Mexico into NAFTA would result in a 31 per cent increase in the current population of the Canada–U.S. FTA (Table 4.1). The three new members of the EC (EC-3) had a combined GNP of 10 per cent of that of the EC-9 while Mexico's GNP is less than 3 per cent of the GNP of Canada–U.S. The EC-3 had a per capita income (PCI) of 43 per cent of the EC-9's, while Mexico's PCI is only 9 per cent of that of Canada–U.S. Thus relatively speaking, the EC-9 added a region almost five times richer than that which the United States and Canada are contemplating. Since market exchange rates tend to undervalue the incomes of less developed countries, an alternative comparison of PCIs using purchasing power parity rates is presented in the second column of Table 4.1. By this measure, the difference in the relative PCIs in the two enlargements is only a factor of two.

Table 4.1   Income and population comparisons

Country	Per capita GNP (1)	Per capita GDP (2)	Population (3)	GNP (1)
Luxembourg	22600	13933	0.4	8.37
West Germany	18530	12604	61.0	1131.27
Denmark	18470	12089	5.1	94.79
France	16080	12190	55.9	898.67
Belgium	14550	11495	9.9	143.56
Netherlands	14530	11468	14.8	214.46
Italy	13320	11741	57.5	765.28
United Kingdom	12800	11982	57.0	730.04
Spain	7740	7406	39.0	301.83
Ireland	7480	6239	3.6	26.75
Greece	4790	5857	10.0	48.04
Portugal	3670	5321	10.2	37.26
EC-9	15134	11994	265.2	4013.19
EC-3	6539	6784	59.2	387.13
EC-3/EC-9	0.43	0.57	0.22	0.10
United States	19780	18339	245.9	4863.67
Canada	16760	16272	26.1	437.47
Canada–U.S.	19490	18141	272.0	5301.15
Mexico	1820	4996	83.6	151.87
Mexico/Canada–U.S.	0.09	0.28	0.31	0.03

Notes:
(1) Data are for 1988, calculated using current exchange rates; data are expressed in U.S. dollars (GNP in billions) and are from the World Bank Atlas.
(2) Based on 1985 PPP exchange rates from the Penn World Table (Mark 5).
(3) Data are for 1988, in millions, and from The World Bank Atlas.

In the extreme hypothetical situation (more extreme than even factor-price equalization) where integration equalized PCI in the core area (EC-9 or Canada–U.S.) and the new addition (EC-3 or Mexico), the PCI of the EC-9 would fall 10.4 per cent and that of Canada–U.S. would fall 21.4 per cent. Thus, by this measure, the addition of Mexico would involve twice the relative adjustment compared with EC enlargement.

4.2.2 Trade Volume Comparisons

In comparing the trade flows of the two regions, 1985 European trade flows and 1989 North American trade flows are used since a comparison of the two cases as they were before integration is desired. In Table 4.2, the size of the bilateral trade flows are compared with each region's total trade; all intra-bloc (that is, intra-core area and intra-EC-3) trade has been netted out. From the core areas' perspectives, the new additions account for a similar proportion of their trade. From the new additions' perspectives, however, the Mexican economy is much more dependent on Canada–U.S. than is the EC-3 on the EC-9. In Table 4.3, the trade flows have been scaled by the GNPs of the regions. Trade with the new additions in both cases is very small in relation to the core area's GNP. Clearly, however, the EC-3 is more important to the EC-9 than is Mexico to Canada–U.S.

Table 4.2   Comparative trade measures

North American FTA	EC Enlargement
$\dfrac{\text{Mex. exports to U.S.–Can.}}{\text{Mex. exports total}} = 81\%$	$\dfrac{\text{EC-3 exports to EC-9}}{\text{EC-3 exports total}} = 53\%$
$\dfrac{\text{Mex. imports from U.S.–Can.}}{\text{Mex. imports total}} = 80\%$	$\dfrac{\text{EC-3 imports from EC-9}}{\text{EC-3 imports total}} = 39\%$
$\dfrac{\text{U.S.–Can. exports to Mex.}}{\text{U.S.–Can. exports total}} = 8\%$	$\dfrac{\text{EC-9 exports to EC-3}}{\text{EC-9 exports total}} = 6\%$
$\dfrac{\text{U.S.–Can. imports from Mex.}}{\text{U.S.–Can. imports total}} = 6.5\%$	$\dfrac{\text{EC-9 imports from EC-3}}{\text{EC-9 imports total}} = 6\%$

Note: All intra-bloc trade excluded. U.S.–Mexico data are from the U.N. and U.S. Dept. of Commerce for 1989; EC data are from the U.N. for 1985.

Table 4.3   Comparative trade GNP measures

North American FTA	EC Enlargement
$\dfrac{\text{Mex. exports to U.S.–Can.}}{\text{U.S.–Can. GNP}} = .5\%$	$\dfrac{\text{EC-3 exports to EC-9}}{\text{EC-9 GNP}} = .8\%$
$\dfrac{\text{U.S.–Can. exports to Mex.}}{\text{Mex. GNP}} = 12\%$	$\dfrac{\text{EC-9 exports to EC-3}}{\text{EC-3 GNP}} = 8\%$
$\dfrac{\text{Mex. exports to U.S.–Can.}}{\text{Mex. GNP}} = 14\%$	$\dfrac{\text{EC-3 exports to EC-9}}{\text{EC-3 GNP}} = 8\%$
$\dfrac{\text{U.S.–Can. exports to Mex.}}{\text{U.S.–Can. GNP}} = .4\%$	$\dfrac{\text{EC-9 exports to EC-3}}{\text{EC-9 GNP}} = .8\%$

Note: All intra-bloc trade is excluded. North American data are from the U.N. and U.S. Dept. of Commerce for 1989; EC data are from the U.N. for 1985.

4.2.3   Tariff Level Comparisons

The tariff levels and changes associated with accession differ significantly in the three European cases; Spanish accession appears to be most comparable to NAFTA. Mexican tariffs are currently similar to Spain's before accession, both being in the 10–20 per cent range. What is also similar and somewhat surprising are the high tariffs which Spain and Mexico have on low-technology, labor-intensive products such as textiles and apparel. Current U.S. tariffs on Mexican products fall mostly between 2 and 4 per cent; this is similar to the EC's tariffs on Spanish goods before accession. Both Spain and the EC had lower tariffs on each others' products relative to third countries; these preferences date back to a preferential trade agreement which was phased in over a period of years beginning in October 1970.

Comparison of the non-tariff barriers is difficult and seems to vary more between the two cases. The EC and the United States are both participants in the Multi-Fiber Agreement, and thus both have quotas on the importation of textiles and apparel. The EC has also maintained significant restrictions on Spanish exports of fruits, vegetables and fat; these quotas are being lifted quite slowly and will not be completely eliminated until 1996 (OECD: Spain, 1990, p.58). The United States has restrictions on Mexican steel, sugar, peanuts and winter fruits and vegetables.

Spain and Mexico also have quotas on numerous products and comparison is difficult. Mexican quotas and licenses primarily restrain U.S. exports of computers, motor vehicles and agricultural products such as grains, dairy products and fruits and vegetables (INFORUM, 1990). Spain had quotas on about one-fourth of imported products and these were especially significant in textiles, apparel, automobiles, color televisions, furniture, plastic products, machinery not elsewhere classified and electrical machinery (Wolcott, 1991; OECD: Spain, 1990). The cases of Portugal and Greece are quite different from that of Spain. Greece has had virtually free access to the EC market since its treaty of association in the late 1960s; however Greece maintained relatively high tariffs on industrial imports, especially on those for which there were Greek manufacturers. Greece had been following a general tendency towards liberalization; import duties and taxes as a percentage of imports fell from 20 per cent in the late 1960s to 10 per cent in 1980 (OECD: Greece, 1990, p. 100). Portugal was one of the original members of the European Free Trade Area (EFTA); beginning in 1977 EFTA and the EC established industrial free trade. Accession did result in the phasing out of some quantitative restrictions and the liberalization of agricultural trade.

4.2.4 Trade Structure Comparisons

Table 4.4 presents a sectoral comparison of the bilateral trade flows of U.S.–Mexico and EC-9–Spain. These nations represent the bulk of the trade flows of their respective blocs and the trade barrier changes in the Spanish case are most similar to those of Mexico. The columns labeled percentage of total trade represent the percentage of total bilateral trade (imports plus exports) for each 1-digit SITC category. For both blocs, SITC 7 – machinery and transport equipment – is the largest sector and represents about 40 per cent of total bilateral trade flows; this sector is more important for U.S.–Mexico than for EC-9–Spain. Both the EC-9, relative to the United States, and Spain relative to Mexico, have more of their exports concentrated in SITC 5 – chemicals and SITC 6 – manufactured articles (classified by material). For both blocs, the core areas have a comparative advantage in chemicals. Manufactured articles are much more important component of Spanish exports than of Mexican exports. Food represents a similar amount of the bilateral trade in the two blocs; it represents a similar

Table 4.4 Comparison of bilateral trade flows by sector

SITC Group	U.S.–Mexico			EC-9–Spain		
	Total trade	U.S. exports	U.S. imports	Total trade	EC-9 exports	EC-9 imports
0. Food and animals	8.6	8.3	9.0	9.2	5.0	13.0
1. Beverages and tobacco	0.5	0.1	1.0	1.3	0.9	1.7
2. Raw materials	4.1	6.2	2.2	6.3	9.7	3.3
3. Mineral fuels	9.7	3.0	15.8	7.5	5.8	9.0
4. Animal, veg. oil	0.3	0.6	0.1	1.0	0.2	1.6
5. Chemicals	5.5	9.1	2.1	11.3	15.5	7.5
6. Manuf. articles	11.0	12.3	9.9	17.5	14.6	20.1
7. Mach. and transport	44.6	44.8	44.4	38.0	40.9	35.4
8. Misc. manufactures	10.3	10.2	10.3	7.8	7.1	8.3
9. Goods not classified	5.3	5.5	5.2	0.2	0.2	0.1

Note: Data for U.S.–Mexico are for 1989 and from the U.S. Dept. of Commerce; data for EC-9 and Spain are for 1985 and from the U.N.

percentage of U.S. and Mexican exports to each other but Spain seems to have a comparative advantage relative to the EC. For the North American region, SITC 3 – mineral fuels – is a much larger sector because of Mexican oil exports. U.S.–Mexico trade also has significantly more SITC 9 – goods not classified by kind.

EC-9 exports are concentrated more in chemicals and manufactured articles, while U.S. exports are more concentrated in machinery and transport, and goods not classified by kind. Spanish exports (or EC-9 imports) are more concentrated in food and animals, chemicals and manufactured articles while Mexican exports (or U.S. imports) are more concentrated in mineral fuels, machinery and transport, and goods not classified by kind. Generally, if an area has a export concentration in a sector, it also has an import concentration in the same sector.

4.2.5 Intra-Industry Comparisons

The degree to which the changes in the trading pattern take the form of intra- or inter-industry trade have implications for both the short-term adjustment costs involved and the long-term effects on wage rates and employment patterns. Generally, intra-industry trade growth is likely to produce fewer adjustment problems and is less likely to alter the distribution of incomes both by factor and by geographical region.

A comparison of the levels of intra-industry bilateral trade calculated at the 3-digit level and summarized by 1-digit SITC category is presented in Table 4.5. The overall level of intra-industry trade and its distribution among the SITC groups are quite similar in the two blocs. Fifty per cent of EC 9–Spain and 52 per cent of U.S.–Mexico trade is intra-industry trade. The level of intra-industry trade within the two blocs is similar and quite high in SITC 6 – manufactured articles and SITC 7 – machinery and transport. There is also significant intra-industry trade in SITC 5 – chemicals – and SITC 8 – miscellaneous manufactures – although it is more significant for U.S.–Mexico in SITC 8 and for Europe in SITC 5. Both blocs have an intermediate level of intra-industry trade in SITC 2 – raw materials with the core areas having significant net exports. There is very little intra-industry trade in either bloc in food products – SITC 0 and 4. The EC has more intra-industry trade in SITC 1 – beverages and tobacco – and SITC 3 – mineral fuels.

Table 4.5 Comparison of intra-industry bilateral trade shares

SITC group	U.S.–Mexico			EC-9–Spain		
	Intra-industry	*Net exports*	*Net imports*	*Intra-industry*	*Net exports*	*Net imports*
0. Food and animals	0.175	0.368	0.457	0.193	0.157	0.650
1. Beverages–tobacco	0.141	0.000	0.859	0.531	0.064	0.405
2. Raw materials	0.403	0.513	0.084	0.340	0.555	0.104
3. Mineral fuels	0.082	0.104	0.814	0.240	0.244	0.516
4. Animal, veg. oil	0.178	0.822	0.000	0.141	0.044	0.815
5. Chemicals	0.370	0.609	0.021	0.573	0.362	0.065
6. Manuf. articles	0.593	0.233	0.174	0.577	0.106	0.318
7. Mach. and transport	0.690	0.133	0.176	0.610	0.202	0.187
8. Misc. manufactures	0.696	0.126	0.178	0.445	0.210	0.345
9. Goods not classified	0.315	0.331	0.354	0.624	0.376	0.000
Total	0.523	0.214	0.263	0.498	0.222	0.279

Notes: EC-9–Spain figures are calculated from 1985 UN data; U.S.–Mexico figures are calculated from 1989 data from the U.S. Dept. of Commerce. Calculated at the three-digit SITC level. Net trade calculated from the core area perspective.

4.2.6 Capital Mobility

Trade liberalizations are likely to have only limited effects in the short term since the capital stock is relatively fixed. In the medium to longer term, however, new investment is able to significantly alter the production structure; access to foreign capital can accelerate this process. Foreign direct investment brings with it technological and managerial improvements which can further alter the comparative cost structure.

Spain historically had a highly restrictive policy towards foreign investment but began liberalization in 1959; by 1985 few limitations remained on foreign direct business and portfolio investments. Controls remained however on short-term capital flows. In the decade prior to accession, Greece generally attempted to stimulate foreign capital inflows by offering numerous incentives, although there were limitations on the remittances of profits and exclusions on investment in the state-controlled and minerals sectors. Foreign investment in Portugal was continuously liberalized after 1976 but had to conform to an investment code that attempted to promote development in priority sectors and increase the transfer of technology.

Mexico has had a long history of restrictions towards foreign investment; beginning in 1971, however, foreign investment was allowed in the Maquiladora sector and has been continuously liberalized. By 1990 there were fairly liberal investment rules if production was primarily for export. Foreign investment for domestic sales in Mexico is still highly restricted; current restrictions include performance requirements, limits on the repatriation of profits and minority ownership requirements.

4.2.7 Labor Migration Patterns

Another interesting parallel between the two blocs concerns the similarities in the migration patterns from the new areas to the core area. Given the wage differences between the core and new areas this is not surprising. In the EC, the free migration of labor was incorporated in Articles 7 and 48 of the Treaty of the European Economic Community and was phased in between 1958 and 1970. Immigrants from the EC-3 accounted for approximately 1.5 per cent of the workforce of the EC-9 in 1984. With the enlargement of the EC, there was concern about increased immigration from the poorer regions and restrictions were placed on immigration from Greece until 1988 and until 1993 for Spain and Portugal. The number of Greek immigrants in the EC peaked in the mid-1970s and then fell slightly during the 1980s; however, with liberalization in 1988, there was an increase in net Greek emigration to the EC-9 (OECD: SOPEMI, 1990). Since 1960 the United States has on average granted permanent residence status to over 50 000 Mexicans a year; in addition, Mexicans may enter as temporary workers – 22 500 Mexicans entered under this provision in 1989. Significant numbers of Mexicans enter the United States illegally; during 1990 over a million Mexicans were caught trying to illegally cross into the United States. Currently about 2 per cent of the U.S. workforce were born in Mexico; this is similar to the proportion of EC-3 workers in the EC-9.

The two cases are therefore similar in that the new regions have a long history of supplying labor to the core areas. One major difference however, is that this trend had decreased or even turned negative in Europe prior to enlargement while Mexican emigration to the United States continues to rise.

4.2.8 Macroeconomic Conditions

The longer-term growth record over the 1965–86 period is similar for all of the new addition countries, with GNP per capita growth of 2.9 per cent for Spain, 3.2 per cent for Portugal, 3.3 per cent for Greece and 2.6 per cent for Mexico. Both Spain (1985) and Mexico (1990) had very high rates of unemployment of slightly over 20 per cent; the unemployment rate in Greece was quite low before its accession in 1981, and only moderately above its historical level in Portugal in 1985.

Real economic growth is more difficult in an environment of inflation. The Mexican case is unique in that none of the European nations have experienced triple-digit inflation as did Mexico as recently as 1987. The Mexican inflation rate is currently (1990) running at about 25 per cent. Greece and Portugal had significant inflation problems also; their rates were about 20 per cent in the years prior to accession while the Spanish rate was only about 10 per cent. In the years prior to accession, Greece and Portugal also had large public sector deficits of over 10 per cent, in Spain they were only half that; for Mexico, the figure for 1988 was 10 per cent (World Bank, 1990, p. 199).

Mexican growth has been seriously constrained by its international debt problems; in 1988 its debt-service to GNP ratio was 8.2 per cent. Greece and Portugal also have significant international debts; the Greek ratio was 3.5 per cent in 1981 and the Portuguese rate was 12.7

per cent in 1985 (World Bank, 1990). Although Spain had a sizable foreign debt, it did not present a significant problem since its international reserve assets were larger than its net debt (Economic Commission for Europe, 1988, p. 76).

The economic situation in the EC-9 (in 1985) and the United States (in 1992) were somewhat similar; both were experiencing stagnant growth. Unemployment in the United States was 50 per cent above its long-term historical level while the EC had unemployment rates in the mid-1980s that were twice their historical levels. In the decade prior to accession, real wage growth for the group most likely to be affected, that is production workers, had been increasing at a significant rate in the EC-9 while wages for similar workers in the United States were falling.

Any alterations in the trade flows produced by trade liberalizations require additional adjustments in the production structure and employment pattern. The question therefore arises as to the similarity in the regions in their ability to adjust to these disturbances in the labor markets. Although the United States is often considered to have labor markets that are more market orientated and thus more flexible than Europe, a major U.N. study found no significant difference between the United States and Western Europe in terms of wage flexibility (Economic Commission for Europe, 1988). Europe is noted for investing more in providing training for the bottom tiers of the labor market.

4.3 INSTITUTIONAL SIMILARITY OF THE TWO INTEGRATIONS

There are significant institutional differences between the proposed NAFTA and EC enlargement. In North America, only a free trade area is being proposed while in Europe an economic union with integrated fiscal, monetary, and regulatory policies is being designed.

4.3.1 Free Trade Area Compared with a Customs Union

Focusing primarily on trade flows, North America will be a free trade area while Europe is a customs union. Mexico will be allowed to maintain its relatively high tariff levels towards the rest of the world while the EC-3 was required to lower theirs to the relatively low levels of the EC Common External Tariff by 1992. This may limit adjustment pressures in Mexico relative to the EC-3; but by creating more trade diversion, this will limit the efficiency gains to Mexico. This situation gives U.S. firms significant competitive advantages relative to producers in the rest of the world that were/are not present for EC firms in the EC-3 market. The tariff situation is similar for the United States and EC-9 in that neither will have to alter significantly their tariff structures towards third nations.

There are some exceptions to this general tendency that have been significant when considering the external aspects of EC integration. Spain and Portugal were required to increase external trade barriers for certain agricultural and food products. The result has been a significant amount of trade diversion for food imports: for Spain in 1985, 78 per cent were imported from non-EC countries but by 1989 only 55 per cent came from non-EC countries (OECD: Spain, 1990).

Since NAFTA will lack a common tariff structure, much more emphasis will have to be placed on rules of origin. This will require a continuation of border checks and fairly complex documentation on the origin of components and so forth; in the EC these transaction costs will be eliminated.

4.3.2 FTA Compared with a Common Market

The EC is in theory a common market, which implies the free mobility of labor and capital. Although NAFTA is being termed a free trade agreement, there are provisions concerning the liberalization of foreign investment in Mexico, and as such NAFTA has some aspects of a common market. Although there will be some liberalization in Mexico's foreign investment restrictions, which should increase capital flows between the two nations, a fully integrated capital market as proposed in Europe is unlikely.

A major difference between the two cases concerns immigration. The EC has allowed the free movement of labor since 1992, while NAFTA will not liberalize immigration except for some minor provisions allowing certain professionals temporary visas. In fact, reducing illegal immigration by assisting in the economic development of Mexico is a significant objective of NAFTA. In both cases there is concern that agricultural trade liberalization will harm small marginal farmers in the new additions, thereby creating a pool of unemployed that will emigrate to the core areas. Actual migration patterns in the two cases are likely to be similar since European enlargement has not significantly increased migration in Europe while NAFTA will not allow increased migration.

4.3.3 FTA Compared with an Economic Union

The EC is not only a common market, it also has significant aspects of an economic union, and with the movement towards the single market of 1992 this is progressively becoming more true. There are numerous programs within the EC that attempt to coordinate economic policies and regulations. For example, the EC-3 had to accept the EC competition policy, limit industrial subsidies, harmonize their tax structures by adopting value added tax and integrate their agricultural systems into the EC's Common Agricultural Policy (CAP). The 1992 program will further harmonize product, safety, environmental and labor standards. Several of these programs, especially the CAP, have a fiscal character which redistributes income between regions of the EC.

The EC-3 have also benefitted substantially from the EC's structural funds, which account for 23 per cent (1991) of the EC's budget. The objective of these funds is to promote economic and social cohesion, which increasingly has meant assistance to backward regions which covers most of the EC-3. These funds, composed of the European Agricultural Guidance and Guarantee Fund, the European Regional Development Fund and the European Social Fund, are used to promote development and structural adjustment and reduce unemployment in the poorer regions and those subject to serious industrial decline. The net transfers of income from the EC-9 to the EC-3 have been quite large; in 1989 the structural funds amounted to 2 per cent of the GDPs of Greece and Portugal and 0.5 per cent of the GDP of Spain (OECD: Portugal, 1992). Even larger transfers are anticipated in future years. In addition, the EC-3 have been able to borrow funds from the European Investment Bank for infrastructure development.

The economic coordination, harmonization and transfers found in the EC will not be part of NAFTA. Each nation in NAFTA will maintain separate national standards, regulations and agricultural programs; there will be no aid similar to the EC's social funds.

4.4 THE CONSEQUENCES OF EC ENLARGEMENT

A full analysis of the effects of EC enlargement is not possible given the short time that has elapsed since accession and the practical difficulty of isolating those developments which were due to EC enlargement. Nevertheless it may be useful to describe generally what has actually happened in the EC in the hope that this can provide some insights into the likely effects of creating NAFTA.

4.4.1 The Effects on the EC-3

The experiences of Spain, Portugal and Greece after accession were significantly different; this complicates the analysis and makes the drawing of implications for NAFTA more difficult. Spain and Portugal appear to have been more successful in capitalizing from EC enlargement than has Greece. For Greece, the decade after accession was characterized by anemic GNP growth, increased unemployment, flat investment and minimal changes in long-term capital inflows and foreign direct investment. For Spain and Portugal, the first five years after accession have been characterized by strong investment-led growth in GNP, due significantly to sizable increases in foreign direct and portfolio investment. To some degree, these differences are due to the fact that the accessions occurred at different points in the European business cycle, however the economy of Greece continued to perform relatively poorly even in the expansionary phase after 1985.

Although there was a significant increase in investment in Spain and Portugal after 1985, these increases should be considered in a longer historical perspective. Investment had fallen in these countries throughout the early 1980s and had reached a low point in 1985. It was not until 1988 that Spanish domestic investment as a percentage of GNP returned to its 1980 level, and Portugal has yet to return to the levels of the early 1980s. Thus much of the increase after 1985 should not be credited with accession but is simply due to the return of investment to its long-term historical level after a cyclical decrease. More generally, the healthy GDP growth of Spain and Portugal during 1986–90 should not be unduly credited with accession; these nations benefited significantly from the economic expansion that occurred throughout Western Europe during this period. Spain's success, relative to Greece and Portugal, in attracting foreign investment was due to the fact that the trade barriers with the EC-9 underwent the most changes and this provided more investment opportunities; also, Spain was more industrialized and had a larger domestic market and a more favorable geographic position.

Accession did not increase the openness of the EC-3 as their trade to GNP ratios remained fairly constant and actually fell for Spain and

Portugal. However, the import penetration ratio (imports/apparent consumption) for manufactured products increased significantly for all three nations. For all three there were significant regional shifts towards the EC-9 in their trading patterns. There is evidence that EC-9–Eastern European trade was diverted to the EC-3 (ECE, 1989). Since the trade liberalizations undertaken by Spain appear to be similar to those that Mexico will undertake, the Spanish case is of particular interest. After four years of EC membership, real (nominal deflated by Laspeyres unit-price indexes) bilateral trade increased by 81 per cent, EC-9 exports to Spain increased by 120 per cent and Spanish exports to the EC-9 increased by 43 per cent. Susan Wolcott (1991) provides a detailed analysis of the resulting Spanish–EC-9 bilateral trade pattern after accession. Several generalizations can be made. First, the sectoral pattern of trade and production was not altered significantly. The largest Spanish import and export industries remained the same, and the industries that had been growing the fastest before unification were the ones that continued to grow the fastest after unification. Second, there was rapid intra-industry trade growth in most sectors. Third, Spanish net exports in almost every industry (and thus the net trade balance) deteriorated; this was the case even in standardized and labor-intensive industries such as textiles and apparel.

An important insight of the Spanish case concerns the degree to which macroeconomic factors dominated the short-term adjustment process. Since there was a much larger decrease in Spanish import barriers from the EC than vice versa, the expectation would be that Spain would experience an incipient trade deficit which would result in an exchange rate depreciation which would in turn improve the competitiveness of Spanish industry (see Krugman, 1990, for a more complicated story). However, Spanish accession significantly increased the desirability of investing in Spain and this (along with higher interest rates from the macro policy mix) created a capital inflow into Spain which put upward pressure on the currency. The net effect was a slight appreciation in the exchange rate and an increase in the net trade deficit (corresponding to the capital inflows).

4.4.2 Effects on the EC-9

Attempting to quantify the effects of EC enlargement on the EC-9 is difficult due to the fact that the changes were relatively small from the perspective of the EC-9. The economic performance of the EC-9 during the 1980s was generally poor, with unemployment rates of twice their historical levels. Although this period corresponded with the EC's enlargement, a connection between the two events is unlikely. In fact, the period after Spanish–Portuguese accession was characterized by falling unemployment and rising real wages, although significant levels of unemployment remained (Bureau of Labor Statistics, 1990, p. 100). The sectoral employment and trade patterns do not provide any breaks in the data that would suggest that enlargement had a negative impact on the EC-9, although it is difficult to estimate what would have happened otherwise.

The impact of EC enlargement on the EC-9 has been limited by several factors. Greece and Portugal had practically free access to the EC-9's industrial markets many years before accession and these adjustment costs had been borne since the early 1970s; thus only Spain's entry entailed potentially significant adjustments. Actual adjustment in the traded-goods sectors was minor since trade expanded in an intra-industry fashion and the EC-9 developed a positive bilateral trade balance. The interest-sensitive sectors might have faced higher interest rates due to the capital outflow but this effect would have been almost imperceptible since the outflow was quite small relative to the overall size of EC-9 investment. Thus the costs to the EC-9 are hard to 'find' because the adjustments were not concentrated in a few import-competing sectors but were diffused throughout an economy which was in the expansionary phase of a business cycle. In the longer run there are still likely to be adjustment pressures in some of the import-sensitive sectors of the EC-9 after the trade balance effects fade and the new investments in Spain and Portugal come online.

4.5 EC ENLARGEMENT AND MEXICO

The enlargement of the EC is a much more comprehensive form of economic integration than what is being contemplated in NAFTA. But the latter is more than just an free trade agreement, since investment, intellectual property, environmental and workers' rights concerns are also being negotiated. As these economies become further integrated, additional areas of economic policy are likely to be coordinated as well. The Mexican case shares similarities as well as dissimilarities with each of the three cases of European enlargement. The pre-accession Spanish case appears to be the most similar with regard to the proposed trade liberalizations and the structure of the trade flows. Since similar trade barrier changes will be applied to similar trade flows, U.S.–Mexican trade should follow the Spanish–EC-9 trade pattern in the immediate years following an agreement. Thus we would expect fairly rapid growth in trade composed of large increases in intra-industry trade, with only limited additional sectoral specialization; Mexico is likely to have a bilateral trade deficit with the United States. Although trade diversion appears to be significant with EC enlargement, it may not be with NAFTA. Mexico already has almost twice the level of trade with the United States that the EC-3 had with the EC-9, thus there is simply less trade to divert. In addition, the agricultural trade diversion which occurred in Europe is unlikely since the United States and Mexico will not integrate their agricultural markets under a unified support program protected by large tariffs. Nevertheless, those nations that compete directly with U.S. or Mexican products will probably suffer some trade diversion. This could be particularly significant for those countries exporting to Mexico if Mexico maintains its large tariffs towards third nations (something the EC-3 could not do).

Given the similarities in trade liberalizations and the likely liberalization of foreign investment in Mexico, the Spanish macroeconomic pattern is also likely to be duplicated in Mexico. To a large degree, the ability of Mexico to duplicate the Spanish experience will depend on whether Mexico can attract sufficient capital inflows. In fact, a major objective of NAFTA for the Mexican government is to resolve their international debt problem by altering investors' expectations so that the capital flight funds, which left Mexico during the 1980s, will return. Mexican growth should benefit, as Spanish growth did, from the large number of unemployed which serve to limit labor bottlenecks.

There appears to be a general long-term global trend within certain income ranges towards convergence in income levels (Maddison, 1987); a major question is whether economic integration promotes this process. More specifically, will NAFTA reduce the income differential between Mexico and the United States and thereby mitigate numerous problems associated with this differential involving trade, the environment and migration? For Spain and Portugal, their GDP per capita increased relative to the EC average in the period after accession; for Greece there has been a significant decrease. The Spanish convergence was due mostly to a reduction in unemployment and an increase in the labor force participation rate, with no convergence in productivity; the Portuguese convergence benefited from all three factors. Although there was some convergence (to the EC average) in Portuguese productivity from 1985–90, convergence was even greater during some other periods, such as from 1975–80. Also, the rate of convergence was quite modest and would require almost half a century to reach even the EC average; much of the increased productivity can be attributed to the structural funds. The European experience therefore suggests that NAFTA will result in very little convergence in U.S.–Mexican productivity; reduced unemployment in Mexico may cause its GDP per capita to converge slightly to U.S. levels.

Perhaps the Greek case provides one of the most important lessons for Mexico: accession in itself provides very little benefit if not accompanied by complementary economic policies. Thus Mexican government policy in establishing a healthy economic climate both at the macro and micro level will remain the key towards Mexican development.

4.6 SUMMARY

A comparison of the similarities between EC enlargement and the addition of Mexico into NAFTA reveals a large number of similarities. Although the income difference between the core areas and the new additions are much greater in the North American case, the two cases are quite similar in terms of numerous ratios of bilateral/total trade, trade/GNP, intra-industry/total trade, the sectoral distribution of trade flows and the structure and level of tariffs.

The economic integration of the EC is much more extensive than what is being planned for NAFTA. The customs union and economic union aspects of the EC are significantly different from a free trade agreement, and the single market of 1992 will further differentiate the two cases. Thus one must be careful in making analogies between the two cases.

In evaluating how enlargement affected the EC, it is difficult to separate out the effects of accession from the business cycle and other long-term trends. It appears that the effects of enlargement on the EC-9 have been marginal; the NAFTA's effects on the United States should be similar. The experience of Greece, Spain and Portugal are quite different, and this limits significantly the number of generalizations that can be made. For Spain and Portugal, there has been growth spurred by investment and capital inflows in the five years after accession. Overall the Spanish case seems most analogous to the Mexican case; however, there are important differences such as their levels of economic development, economic sizes, and general levels of macroeconomic stability. If the Spanish experience is duplicated we would expect Mexican growth to increase, an increased trade deficit for Mexico matched by capital inflows from the United States, significant increases in the volume of bilateral trade, with much of it being intra-industry, some trade diversion and a very small convergence in incomes and productivity.

5 The Linking Giant: An Analysis and Policy Implications of the Canada–Mexico–U.S. Free Trade Area

M. Reza Vaghefi

5.1 INTRODUCTION

Since January 1989, when the Canada–U.S. Free Trade Agreement was signed, the global economic landscape has witnessed dramatic changes. With the collapse of Eastern Europe and communism and the emerging economic powerhouse in the Far East, Mexico became interested in the Canada–U.S. negotiations and there are prospects of a new economic alliance in North America. There are potential ingredients in the emerging North American Free Trade Agreement (NAFTA) for a North American common market. This chapter examines the potential of, and possible problems on the way to, the achievement of NAFTA.

Negotiating the elimination of trade barriers between North American countries has proven to be a gauntlet of expected and unexpected complications. The negotiations are unique. Although other nations have successfully reached trade agreements, these negotiations provide the first opportunity to bring together such disparate economies. Canada with a 1991 GDP of $501 billion, Mexico, with $283 billion, and the United States, with $5673 billion (a total of $6457 billion), are, economically, very different nations. The largest industrial nation negotiating free trade with an industrial country and a developing country provides a unique opportunity to contrast the complex difficulties associated with each. Never before has a developing

*I am indebted to Stanley Jurewicz for his contribution to an earlier draft of this chapter. Historical development of this topic has been presented in volume III, no. 4 of the *International Trade Journal*, Summer 1989.

country like Mexico entered negotiations with an industrial power with the intent of achieving real free trade. The difficulties influencing both negotiations and the manner in which they were overcome, or not overcome, provides valuable lessons for future negotiations.

5.2 THE THEORETICAL REASONING

Fact, experience and intellectual reasoning support the proposition that free trade will ultimately benefit all nations of the world. While some may be bigger winners than others, all nations will be winners compared with their present positions. Free trade will result in the optimum outcome for all nations. Greater access, application of economy of scale and rationalization of industries provide lucrative incentives for the free movement of capital, labor and entrepreneurship. Geographical proximity, empirical evidence and simple observation of trade pattern leave little doubt about the wisdom of a free trade area in North America.

5.2.1 The Trilateral Benefits

Tables 5.1, 5.2, and 5.3, which cover 15 years of economic activities, provide the most solid evidence of one unquestionable fact, that is the heavy dependence of Canadian and Mexican economies on the U.S. market. The United States has depended on Canada and Mexico, on average, for 26 per cent of its exports over 15 years, while 74 per cent of its exports went to the rest of the world, which included Japan (10 per cent) and Europe (20 per cent), with the remainder going to Middle East and other countries in Southeast Asia and Africa.

While the United State's dependence on Canada and Mexico does not seem overwhelmingly astonishing, Table 5.2 provides a clear picture of Canada's increasing dependence on the U.S. market and minimal dependence on the Mexican economy. Canada depends on the rest of the world for about 30 per cent of its exports.

Mexico's reliance on the U.S. market almost matches that of Canada. Table 5.3 indicates that Mexico depends on the U.S. market for two-thirds of its exports, and this dependence has been on the rise during normal years. Of the three members of NAFTA, Canada's dependence on export is the highest, ranging from 20 per cent to 26 per cent of GNP. All these data exhibit a de facto integration of the three

64

Table 5.1 U.S. exports (in millions of dollars)

	To Canada	%	To Mexico	%	Rest of World	%	Total	U.S. GNP	% of Export
1976	24109	20.89	4990	4.32	86314	74.78	115413	1782800	6.473
1977	25788	21.26	4821	3.97	90697	74.76	121306	1990500	6.094
1978	28372	19.74	6680	4.65	108710	75.61	143762	2249700	6.390
1979	33096	18.18	9858	5.42	139053	76.39	182007	2508200	7.256
1980	35395	16.03	15146	6.86	170240	77.10	220781	2732000	8.081
1981	39564	16.93	17789	7.61	176385	75.46	233738	3052600	7.657
1982	33720	15.89	11817	5.57	166737	78.54	212274	3166000	6.704
1983	38244	19.07	9082	4.53	153202	76.39	200528	3405700	5.888
1984	46524	21.35	11992	5.50	159373	73.14	217889	3772200	5.776
1985	47251	22.17	13635	6.4	152260	71.43	213146	4040300	5.314
1986	45333	20.86	12392	5.70	159567	73.43	217292	4235000	5.130
1987	59814	23.65	14582	5.77	178488	70.58	252884	4488000	5.634
1988	69484	21.78	20473	6.40	229795	71.82	319952	4873700	6.56
1989	78266	21.51	24969	6.86	260.572	71.62	363807	5200000	6.9
1990	82959	21.10	28375	7.22	281772	71.68	393106	5465200	7.6
Avg		20.03		5.79		74.18	100		6.97

Source: Trade data from International Monetary Fund, *Direction of Trade Statistics*, various years.

Table 5.2 Canada's exports (in millions of dollars)

	To U.S.	%	To Mexico	%	Rest of World	%	Total	Canada GNP	% of Export
1976	26262	64.69	219	0.54	14117	34.77	40598	167200	24.28
1977	29339	67.37	206	0.47	14001	32.15	43546	181000	24.05
1978	32624	67.36	203	0.42	15602	32.21	48429	200400	24.16
1979	37658	64.60	207	0.36	20433	35.04	58298	224000	26.02
1980	41068	60.63	419	0.62	26243	38.74	67730	254600	26.60
1981	46454	63.98	609	0.84	25549	35.18	76612	284200	25.54
1982	46529	65.41	369	0.52	24232	34.06	71130	290500	24.48
1983	53848	70.16	310	0.40	22587	29.43	76745	313800	24.45
1984	66300	73.43	278	0.31	23715	26.26	90293	340800	26.49
1985	68283	75.22	287	0.32	22210	24.46	90780	353600	25.67
1986	67183	74.89	288	0.32	22235	24.78	89706	374000	23.98
1987	71455	72.84	398	0.41	26251	26.75	98104	402100	24.39
1988	81962	70.31	404	0.35	33996	29.34	116571	586190	19.89
1989	85305	70.69	525	0.44	34843	28.87	120673	627960	19.22
1990	95388	72.66	488	0.37	35402	26.97	131278	647620	20.27
Avg		68.95		0.45		30.6	100		23.97

Source: Trade data from International Monetary Fund, Direction of Trade Statistics, various years.

Table 5.3 Mexico's exports (in millions of dollars)

	To U.S.	%	To Canada	%	Rest of World	Total	Mexico GNP	% of Export
1976	2111	60.85	48	1.38	1310	3469	66100	5.25
1977	2378	65.54	45	1.08	1388	4171	72600	5.75
1978	4057	68.14	62	1.04	1835	5854	92450	6.44
1979	6252	69.60	75	0.83	2656	8983	110000	8.17
1980	10072	64.74	117	0.75	5368	15557	128000	12.15
1981	10716	55.29	661	3.41	8004	19381	150000	12.92
1982	11887	56.17	579	2.74	8697	21163	162000	13.06
1983	13034	58.41	467	2.09	8812	22313	158000	14.12
1984	14130	57.95	495	2.03	9757	24382	161300	15.12
1985	13341	60.35	393	1.78	8371	22105	173400	17.75
1986	10424	64.67	224	1.39	5472	16120	122700	13.14
1987	13265	64.63	312	1.52	7909	20526	166700	12.31
1988	13419	63.05	316	1.52	7864	20765	17400	11.93
1989	16092	70.04	272	1.18	8475	22975	204000	11.26
1990	21922	73.12	725	2.42	7335	29982	—	
Avg		65.42		1.4	33.18	100		11.38

Source: Trade data from International Monetary Fund, Direction of Trade Statistics, various years.

markets, with the United States providing the bulk of integrating glue with a higher purchasing power of $22.40 per capita income vis-à-vis Canadian's $21.98 and Mexico's $3.40. The United States, while providing a higher buying capacity and a powerful economic engine, will be able to benefit from increased exports to Mexico in the short term and more investment income from Canada in the long term.

5.2.2 Fundamental Premises of NAFTA

NAFTA's basic objectives have been the following:

- Promoting trade liberalization.
- Improving the climate for trilateral investment.
- Resolving problems arising from disputes over the use of subsidies and countervailing duties.
- Creating new rules to govern trade in services and liberalize the financial services market.
- Promoting multilateral cooperation on trade and investment issues in the General Agreement on Tariffs and Trade (GATT).

Econometric estimates of the effect of the free trade agreement indicate that, for Canada, NAFTA will push real income to a higher level by 2.5 per cent to 3.5 per cent of GNP; estimates for the United States are much smaller, and considerably less than 1 per cent of GNP.[1] The disproportionate effect is understandable, given the fact that Canada will obtain better access to a market that is more than ten times its size, and U.S. firms will gain only a small amount relative to their already large home market.

5.3 CANADIAN AND MEXICAN BARRIERS TO TRADE NEGOTIATIONS

European Community members have negotiated among themselves as relatively economic equals, at least at the inception in 1957. The North American countries entered negotiations with very different perspectives. The most critical barriers in Canada and Mexico are based on their being semi-industrial and developing countries. Their fears regarding their dealings with the United States range from resentment in Canada to xenophobia in Mexico.

5.3.1 Nationalism

Canada, with an economy and population one-tenth the size of the United States, has a part of its population that worries 'that their country would be swallowed whole, culturally and economically' by the United States.[2] This sensitivity to the issue of Canadian sovereignty has historically been a hindrance to U.S.–Canadian trade negotiations. When, in 1911, speaker-designate Champ Clark of the U.S. House of Representatives expressed his hope to see the U.S. flag flying over Canada, the ruling party in Canada, which had been undertaking serious trade negotiations with the United States, was ousted.[3] This issue is still powerful and is being used by provincial governments in Ontario and elsewhere in Canada. Shirley Carr, president of the Canadian Labor Congress, claimed, when the latest free trade negotiations were in progress, 'It is in the interest of the United States to try and take over Canada. It has always been, ever since Canada was first formed . . . They want to disrupt and disturb everything we have and bring us down to their level.'[4] That attitude has historical roots going back at least 110 years when Canadians, led by conservatives, decided to erect a huge tariff wall, effectively thwarting the entry of manufactured goods from the United States.

The tariff made living more expensive in Canada but also allowed indigenous companies to emerge, albeit inefficient ones. But what a liberal government could not achieve over 100 years ago a conservative government made possible, realizing a huge Canadian dependence on the U.S. market, by entering into agreement, and thus NAFTA took shape. Although a general ambivalence is felt by many Canadians and this generally shows up during bad times, Canada is not as xenophobic as Mexico.

Whereas the basis for Canadian animosity is ambiguous, Mexican distrust is bolstered by real and frequently tragic historical events. The Treaty of Guadalupe, which ended the U.S.–Mexican War in 1848, not only confirmed the U.S. claim to disputed Texan territory but annexed that part of Mexico which is today known as the states of New Mexico and California, fully half of Mexico's land. Mexican textbooks, folk songs and shrines still consecrate the Mexican heroes of that war. In 1913 the U.S. ambassador to Mexico helped organize a rebellion that overthrew the Mexican government. Today that event is recounted as the *decena tragica* (ten tragic days). U.S. military intervention at Veracruz in 1913 and their pursuit of Pancho Villa in 1916 are important history lessons to Mexican school children. These events

have led to what one U.S. historian calls a 'virulent, almost pathological, Yankeephobia'.[5]
   The anti-U.S. sentiment in Mexico and Canada at worst inflames nationalism in response to perceived threats to national sovereignty. At best, it reinforces the fear in both countries that they will be coerced into accepting the outcome least favorable to them during trade negotiations, and the new 'world order', benign as it seems, may not have minimized those fears. But the reality is that the trade data in Tables 5.1, 5.2, and 5.3 produce a different scenario.

5.3.2 Dependence

A barrier which has frequently arisen to thwart trade negotiations with the United States has been the asymmetric dependence the Canadian and Mexican economies have on the U.S. economy. The United States and Canada are the world's largest trading partners. While this partnership is very important to the United States, it is most critical to Canada and Mexico. Canada has made many attempts to decrease this dependence. John Diefenbaker in the 1950s and Pierre Eliot Trudeau in the 1970s and early 1980s were both elected prime minister in no small part because of pledges to increase trade with Europe. Both failed to keep their word. Canadian exports to Europe dropped from 19 per cent in 1970 to 7 per cent in 1985.[6] Canadian attempts to decrease its dependence on the United States highlight the concern about overdependence on their southern neighbor.
   Analysis of foreign trade data of the NAFTA members (Figures 5.1, 5.2 and 5.3) provides an unambiguous pattern that may indeed have provided the guiding light in choosing the policy alternatives which ultimately led to the Canada–U.S. agreement, and now to Mexico's decision to join the emerging process, which has a unique potential unrivalled in global trading systems.
   Frustrated Canadian attempts to divert trade away from the United States frequently, and grudgingly, reverted Canadian attention instead to increasing trade with the United States. While wishing to diversify economic ties, Canada was receptive to economic reality. Mexico has been more recalcitrant in its disdain of economic dependence on the United States. Mexico, the third largest trading partner of the United States, sends on average 65 per cent of its total exports to the United States, while only 5.8 per cent of U.S. exports since the late 1970s have gone to Mexico. Clearly the Mexican economy is strongly dependent on the U.S. economy – Mexico is unhappy about this dependence and

Figure 5.1   United States exports and imports, 1980 and 1990

United States exports

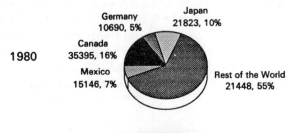

1980

Germany
10690, 5%

Japan
21823, 10%

Canada
35395, 16%

Mexico
15146, 7%

Rest of the World
21448, 55%

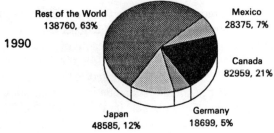

Rest of the World
138760, 63%

1990

Mexico
28375, 7%

Canada
82959, 21%

Japan
48585, 12%

Germany
18699, 5%

United States imports

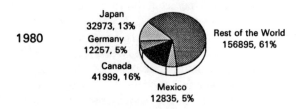

1980

Japan
32973, 13%

Germany
12257, 5%

Canada
41999, 16%

Mexico
12835, 5%

Rest of the World
156895, 61%

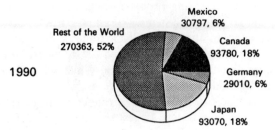

Mexico
30797, 6%

Rest of the World
270363, 52%

Canada
93780, 18%

1990

Germany
29010, 6%

Japan
93070, 18%

Note: Data in millions of dollars

Source: 1991, International Monetary Fund, *Direction of Trade Statistics*.

Figure 5.2   Canada's exports and imports, 1980 and 1990

Canada's exports

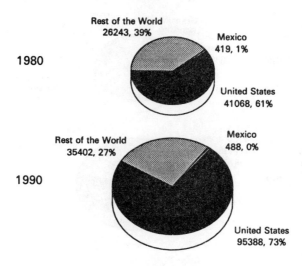

1980

Rest of the World
26243, 39%

Mexico
419, 1%

United States
41068, 61%

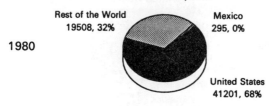

1990

Rest of the World
35402, 27%

Mexico
488, 0%

United States
95388, 73%

Canada's imports

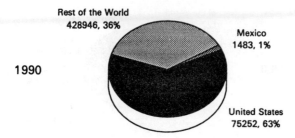

1980

Rest of the World
19508, 32%

Mexico
295, 0%

United States
41201, 68%

1990

Rest of the World
428946, 36%

Mexico
1483, 1%

United States
75252, 63%

Note: Data in millions of dollars
Source: 1991, International Monetary Fund, *Direction of Trade Statistics*.

Figure 5.3   Mexico's exports and imports, 1980 and 1990

Mexico's exports

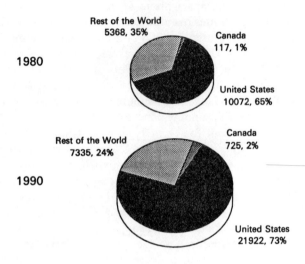

1980

Rest of the World
5368, 35%

Canada
117, 1%

United States
10072, 65%

1990

Rest of the World
7335, 24%

Canada
725, 2%

United States
21922, 73%

Mexico's imports

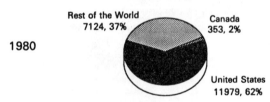

1980

Rest of the World
7124, 37%

Canada
353, 2%

United States
11979, 62%

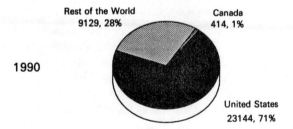

1990

Rest of the World
9129, 28%

Canada
414, 1%

United States
23144, 71%

Note: Data in millions of dollars
Source: 1991, International Monetary Fund, *Direction of Trade Statistics*.

fears that a free trade agreement would solidify its 'dependent–
hegemonic relationship'.[7]
Since the late 1970s, on average 26 per cent of U.S. exports went to
its northern and southern neighbors (Table 5.1), whereas Canada's
exports to the United States over the same period constituted about 70
per cent of its total exports (Table 5.2). Mexico has become increasingly dependent on the U.S. market, on average shipping 65 per cent of
its total exports to the United States. Between 1976 and 1990 alone,
Mexican exports to the United States increased by almost 1000 per cent
(Table 5.3). While the Canadian economy is heavily dependent on the
wellbeing of the U.S. economy, one estimate claims that Mexican
exports fall by 2 per cent for every 1 per cent reduced growth in the
U.S. economy. Recession in the United States is extremely detrimental
to Mexico. Mexican business consultant and former finance ministry
official, Josue Saenz, related the feelings of many Mexican businessmen
toward a free trade agreement: 'They say Mexico is a sardine, and the
U.S. a shark. And if they swim in the same ocean, the shark will eat the
sardine'.[8] Mexico responded to this fear of dependency by nationaliz-
ing industries, subsidizing import substitution, raising trade barriers
and limiting foreign investment. Reduction of dependence on the
United States has been a major consideration in shaping Mexican
economic and foreign policy since the early 1940s. A recent example
was the decree by the Lopez Portillo administration that no more than
50 per cent of Mexican oil exports could go to the United States.[9]

5.3.3 Competitive Advantage

Traditionally, resistance to free trade in Canada has been led by
Ontario industrialists trying to retain their tariff protection. Western
provinces have advocated free trade as a means of obtaining cheap
manufactures for their resource-intensive industries. Tariffs have
enabled Canadian manufacturers to compete with U.S. imports in
the Canadian market. Free trade would force Canada to rationalize its
industries in order to compete with U.S. manufacturers. This process
would force Canadian industry to become more specialized and to
realign its manufacturing base, with plant closures and business failures
following, although access to the U.S. market would ultimately be
beneficial to Canada.

The foremost concern of Mexico regarding competitive advantage is
the same as that of Canada, but greatly magnified. With a GNP one-
twentyfifth the size of that of the United States, Mexican industry is

not yet able to achieve economies of scale. Mexico's earlier inward-looking industrial policy resulted in an industrial base which was heavily subsidized and uncompetitive by international standards. As a result, many Mexican industrialists, with the exception of those who already enjoyed footholds in the United States, resisted a free trade agreement and the competition it would unleash.[10]

5.4   OVERCOMING THE BARRIERS TO FREE TRADE

Mexico and Canada have attempted to manipulate their relations with the United States in order to offset their less powerful positions. Due to its position of power, the United States had to wait for Mexico and Canada to initiate free trade negotiations – any attempt on the part of the United States to initiate such negotiations would inevitably have been seen as attempts to exploit its advantageous position. The barriers to free trade negotiations described above were powerful deterrents to Canadian and Mexican initiatives. Yet Canada and, later, Mexico did initiate negotiations. Why and how they overcame strong protectionist tendencies provides some interesting insights into global economic positioning.

5.4.1   World Economic Changes

The fact that the world economy is rapidly changing is recognized in both Canada and Mexico. The international environment is becoming increasingly competitive. With the emergence of Europe–1992 and serious attempts to enlarge the EC from 12 members to 17 or more with a population of 380 million, and increasing cooperation among East Asian nations enjoying a $4044.5 billion GNP and a population of over 488 million (excluding China), Canada was facing the prospect of being one of the world's only industrialized nations without free access to a market that has high purchasing power and at least 250 million consumers. Although an industrially competent country, the lack of a large market would have prevented Canada from achieving the economies of scale enjoyed by its international competitors, and this competitive disadvantage would have been devastating to the Canadian economy.

Mexico is equally aware of the increasingly competitive international environment. In order to be able to compete, Mexico realizes it must modernize its industrial base and such modernization will require additional foreign investment. While Mexico would prefer to entice foreign investment from sources other than the United States, its attempts to do so, with the exception of Maquiladora, have been thwarted. Critical events in Europe have motivated Mexico to seek U.S. investment. The emergence of the EC and the upheaval in Eastern Europe relegated Latin America to the bottom of European considerations. After touring European capitals in early 1990, Mexican President Carlos Salinas became acutely aware that the foreign investment and aid he was seeking was being funnelled into Eastern European development and Western European preparations for integration in 1992. The experience impressed him enough to shortly thereafter announce the privatization of commercial banks and his intention for Mexico to seek free trade. This was a dramatic change of direction. In September 1989, a U.S. trade official was reported in the *Washington Times* as having commented that 'the Salinas government has rejected the idea [for a free trade agreement] at every opportunity'.[11]

5.4.2 Fear

A former Canadian prime minister said, 'Living next to the United States was somewhat similar to sleeping next to an elephant. No matter how good-natured the beast, one does tend to notice every grunt and twitch.'[12] Canada, a country whose exports were increasingly represented by manufactured products rather than raw materials, was attuned to the protectionist rhetoric in the United States Congress that was growing as fast as the U.S. trade deficit. The General Agreement on Tariff and Trade (GATT) negotiations apparently would not be enough to protect Canada. By entering bilateral free trade negotiations, Canada hoped to preempt the protectionist threat and ensure itself access to the U.S. market.

Canada was finding it increasingly difficult to deal with U.S. red tape. For example, while only 30 per cent of the countervailing duty and anti-dumping cases filed against Canada with the International Trade Commission (a procedure established under GATT) by U.S. corporations were substantiated, the process routinely took up to a year to complete. U.S. corporations were using this mechanism as a weapon to inflict considerable damage and expense on Canadian rivals.[13] Canada was motivated to negotiate a free trade agreement to allow it more leverage in such disputes with the United States.

Mexico, while fearing increased U.S. protectionism, had additional fears following the successful conclusion of the Canada–U.S. FTA. A study by the Mexican Ministry of Commerce found that the United States and Canada were likely to purchase products from each other that were previously purchased from Mexico.[14] A further impetus was provided by President Bush's announcement of his Enterprise for the Americas. By announcing its intention to negotiate free trade with the United States, Mexico has differentiated itself from the rest of Latin and South America. It hopes, by doing so, to deal with the United States on its own terms rather than on the same terms as other Latin American countries.[15]

5.4.3 Changes in Mexico

Mexico had to develop the confidence to compete with a highly industrialized country without being exploited. Numerous changes have built that confidence.

The Mexican population is becoming increasingly educated. Mexico's literacy rate is over 87 per cent, comparable with that of the United States (99 per cent), and the number of university graduates is steadily increasing.[16] Mexican universities graduate more engineers per capita than the United States. Managers and engineers trained in the United States are becoming more common in Mexican industry and have been assuming leadership positions. IBM showed its confidence in educated Mexicans by investing $30 million in postgraduate training for Mexican engineers and computer professionals. Mexico is developing a highly competitive and increasingly technical workforce.[17]

Mexican workers are also becoming increasingly productive and Mexican industries increasingly diverse. In 1980 two-thirds of Mexican export income was from petroleum-related products. In 1990 it was only one-third. Maquiladora assembly facilities are no longer seen as havens for low cost, low quality production. Rather they are developing a reputation for being efficient producers complex goods.[18] When one considers that hourly compensation in manufacturing in Mexico is $1.80, in United States $14.77 and in Canada $16.02, one is not surprised at the rapid growth of investment by American corporations moving into Mexico.[19] Rates of productivity growth in U.S. subsidiaries in Mexico exceed similar rates in the United States. Ford's Hermosillo assembly plant is equal in quality to the best assembly plant in the world, run by Daimler-Benz. Management techniques in some Mexican companies are of the latest variety. Vitro Corporation, a glass producer, is able to compete internationally because of the high quality of its products rather than lower prices. Its engineers frequently visit customers, and product development teams bring together members from different parts of the organization.[20] Mexico is becoming increasingly capable of competing internationally in other than labor-intensive industries.

The last facet in building its confidence was the alleviation of the fear that all foreign investment was merely an attempt to exploit Mexico. The success of the maquiladoras and the failure of its own protected industries to adequately provide for the domestic market has served to highlight the benefits provided by foreign investment. Additionally, Mexico's young population, 40 per cent of whom are under 15, is becoming increasingly able to look to the future rather than to the past. Being able to deal with the future rather than reliving the past is perhaps the most important, and difficult, adjustment Mexico has made.

5.4.4 One Key Player

In Canada and Mexico free trade negotiations could not have been initiated without a strong leader who was prepared to take risks. The barriers discussed in this chapter have made it very risky for the leader of either country to initiate such negotiations – in Canada, Prime Minister Brian Mulroney risked his political career by initiating free trade negotiations with the United States. His campaign for reelection in 1988 was really a referendum fought by free trade proponents and opponents.[21] His successful reelection was confirmation that Canadians were ready for free trade.

Mexican President Carlos Salinas faced far greater obstacles than did Prime Minister Mulroney. Carlos Salinas had to establish himself as a powerful and capable leader before being in a position to initiate free trade negotiations. A weaker leader would not have been able to overcome the barriers to such an initiative. Salinas entered office with an economy in shambles. The steps he took to turn the economy around broke with long-standing Mexican tradition: the negotiation of a debt-reduction agreement, the sale of government-owned businesses, the reduction of tariffs and the liberalizing of foreign investment regulations. Strong-arm tactics were used to dislodge long-pampered Mexican industrialists and politicians. The first year in office he used the army to arrest the corrupt yet powerful oil workers' union boss, Joaquin Hernandez Galicia. He broke labor strikes and privatized business.[22] Despite the hardship such actions have caused, Salinas is still extremely popular with the Mexican populace. Only with a leader able to overcome opposition entrenched through years of mismanagement and corruption could Mexico be in a position to overcome the barriers to a free trade initiative.

5.5   U.S. BARRIERS AND MOTIVATIONS

5.5.1   Barriers

The United States faced far different barriers to entering free trade negotiations than either Canada or Mexico. In its position as the more powerful nation it was more likely to achieve its own best outcome without free trade pacts. This was especially true when considering free trade with Mexico and it is by no means coincidental that Canada, much more the equal of the United States, was the first of the two nations with which the United States negotiated. An additional barrier, based on its position of relative power, is that any attempt by the United States to initiate the negotiations would have been immediately suspect in the other country. This latter barrier was overcome when Canada and Mexico initiated the negotiations.

Another barrier to U.S. participation in free trade negotiations was the U.S. Congress and the special interest groups it represents. Prior to the negotiations, industries that currently enjoyed the benefits of trade protection, known to economists as subsidized industries, lobbied their Congressmen and rallied the public to apply pressure to preempt the negotiations. When negotiations are entered into, negotiators on both sides can expect to see their reforms 'clawed back' in the ratification process, and this happened to some of the provisions of the Canada–U.S. FTA.

The initiation of negotiations with Mexico is facing intense opposition and special interest groups are threatening to undermine the start of the negotiations. Within days of President Salinas' announcement of his desire to negotiate an agreement, the AFL–CIO was urging Congress to reject the proposition.[23] Agriculturists and the textile industry have been equally aggressive in their opposition to negotiations with Mexico. The Florida agricultural commissioner, Doyle Conner, responded to the proposal by saying, 'The economic ramifications for Florida growers would be disastrous if we entered a free trade agreement.'[24] Ironically, while Mexico fears that it will be unable to compete with a high-wage, highly industrialized country, U.S. critics worry about competing with Mexican labor costs, which are as low as

$3.50 a day. The battle in Congress promises to be heated and frequently irrational.

5.5.2 Motivation

The United States, being the world's largest economic power, would prefer a world-encompassing free trade agreement. Ironically, it was its frustration with the GATT negotiations that gave the United States its primary impetus to negotiate with Canada. While a proliferation of bilateral and regional free trade agreements may ultimately result in numerous 'trading blocs' competing between each other, the initial purpose of the Canada–U.S. FTA was to save GATT. Former U.S. trade representative William E. Brock 'floated the idea to save GATT'. He said, 'The idea came after one of the more miserable experiences I've gone through, trying to win agreement among GATT nations on ways to keep the system from self-destructing, to keep it from falling apart at the seams'.[25] In response the United States entered negotiations with Canada. Its objective was two fold. First, it was a threat to intransigent GATT nations – if the United States displayed its willingness to enter bilateral agreements it was hoped other GATT nations would be motivated to negotiate more seriously. Second, the United States hoped that the Canada–U.S. FTA would set precedents for GATT negotiations in the development of rules on trade in services and trade-related investment.[26]

The United States' motivation for entering into negotiations with Mexico is less clear. Proponents counter the loss of U.S. jobs argument by the reasoning that losing jobs to Mexico is better than losing jobs to overseas companies. Herminio Blanco, a University of Chicago economist said, 'Some industries in the United States are declining. They are going down with or without a free-trade agreement. What the U.S. must decide is where do they want the jobs to go – to Mexico or to somewhere else'.[27] By directing those jobs to Mexico the United States gains the benefits of an improved Mexican economy. Political stability, a decrease in illegal immigration and access to an economy with improved buying power are some of the benefits mentioned. Others propose that the agreement would create jobs in U.S. industries that support trade and commerce.[28]

Another motivation may be to stimulate the GATT negotiations while at the same time setting the stage for a Western Hemisphere free trade region, should the global negotiations fail. Already Canada has become involved with the U.S.–Mexican proposal and a North

American Free Trade Area is becoming more likely. Furthermore, Mexico could become the linchpin to President Bush's 'Enterprise of the Americas'. An agreement between Mexico and the United States may encourage other Latin American countries to seek entry into the agreement. If Mexico were to gain special access to the U.S. market, countries such as Brazil would find it in their own self-interest to seek similar access to avoid competitive disadvantage. An area of free trade and investment stretching from Alaska to Tierra del Fuego would provide a formidable trading block.

5.6 FROM NORTH AMERICAN FREE TRADE AREA TO NORTH AMERICAN COMMON MARKET

One motivation for the United States to open trade negotiations is to provide a counterbalance to other free trade regions should global negotiations prove inadequate. With Canada and the United States already working together, the question arises as to whether incorporating Mexico into the agreement would provide any more significant weight to a North American counterbalance.

A trilateral free trade pact in North America promises, like a hub and spoke system with the United States as the hub and Canada and Mexico as spokes, to unite a population of over 350 million people in the free flow of goods and investment. Despite all the ramifications it would entail, it appears that Mexico is becoming more likely to support it. Francisco Garza, president of Monterrey-based Vitro Industries Basicas says, 'I think the attitude today in Mexico is "Let's just do it." There's going to be pain. There's going to be injustices. But in the end there will be more benefits overall.' Salinas claims that the more than $6 trillion dollar manufacturing and consuming bloc that would result would be fully equivalent to post-1992 Europe and East Asia.

Mexico appears to be an equal candidate, if not a superior one, for stimulating economic expansion than are the Eastern European countries that Western Europe is currently so enthralled with. Mexico's international debt burden, while large, is less of a burden than the debt faced by many Eastern European countries. Austerity measures recently and renegotiations of payments have allowed Mexico to reduce its international transfers due to debt from 6 per cent of GNP in the 1980s to 2 per cent of GNP in the early 1990s. The per capita burden this debt places on Mexicans is half that placed on Hungarians and 30 per cent less than that shouldered by Poles, not to mention the $60–70 billion owed by the Soviet Union. Inflation has been driven down from 167 per cent in 1987 to 20 per cent in the early 1990s and it promises to go lower. In 1991 Polish inflation was 1000 per cent while Yugoslavia suffered 250 per cent inflation. The populace of Mexico boasts a literacy rate that is comparable with that found in Eastern Europe. The Eastern Europeans, however, have suffered through more than 40 years of communist rule that has severely reduced their motivation to work and depleted the supply of managerial talent. The market potential in Mexico appears to be far superior to that of Eastern Europe, at least for the rest of twentieth century.

The entry of Mexico into a trilateral free trade accord would have much greater ramifications for the United States than the bilateral agreement now in effect with Canada. The reasons cited above and the very low-cost labor combine to make Mexico a nation that would be much more competitive with the United States than Canada is. The greatest benefit for the United States would be the rationalization and stimulation such competition would provide. While the agitation such a pact would release in certain American industries as the economy realigned itself with the new reality would be apparent, it would ultimately be beneficial. U.S. industry would be forced to become more competitive. The shifting of certain industries to lower factor cost nations is inevitable. With such a move could come some experimentation with competition in service sectors, which may take a few years.

Canada did not initiate trade talks with Mexico, but once Salinas had embarked on his course they decided not to be left out and so joined the negotiations. In the past, Mexico has not figured prominently among Canada's political, economic or strategic priorities. This is evident from export data (Tables 5.2 and 5.3) which show less than 0.5 per cent for Canadian exports to Mexico and 1.4 per cent for Mexican exports to Canada since the late 1970s. But it is the promise that Mexico presents that entices Canada to be more than a jubilant observer of negotiations between the United States and Mexico. Although a recent study of the U.S. International Trade Commission concluded that a U.S.–Mexico agreement could dilute the benefits of the Canada–U.S. FTA by making Mexican products more competitive than Canadian ones in the U.S. market,[29] it is precisely the benefit of a broader market that would allow people to benefit from economic expansion. The marriage of U.S. capital and technology with Mexican labor and Canadian natural resources has the potential to create a formidable challenge to EC and Pacific Rim markets. Short-term problems, such as recent squabbles over Honda parts, subsidized

Canadian lumber, or U.S. brewers' free entry into heavily protected Canadian markets, have to be resolved for the common good. Down the road one can almost see a North American common market – it took Europeans 35 years to achieve a barrier-free market; it seems that period can be shortened to 20 years in North America.

REFERENCES

1. Canadian Department of Finance, 1988.
2. Erik Gunn, 'Canada Still Leery of Free-Trade Pact', *Milwaukee Journal*, 18 March 1990.
3. Paul Wonnacott, *The United States and Canada: The Quest for Free Trade* (Washington, D.C.: Institute for International Economics, 1987) p. 14.
4. Ibid., p. 15.
5. Sidney Weintraub, *Free Trade Between Mexico and The United States?* (Washington, D.C.: The Brookings Institute, 1984) p. 15.
6. Paul Wonnacott, p. 16.
7. Sidney Weintraub, p. 2.
8. Debra Beach, 'Mexico Pins Hopes on Free-Trade Pact', *Houston Chronicle*, 26 June 1990.
9. Sidney Weintraub, p. 16.
10. 'Enormously Consequential', *Forbes*, April 1980.
11. Karen Riley, 'Free-Trade Pact With Mexico Advocated', *Washington (D.C.) Times*, 19 October 1989.
12. Paul Wonnacott, p. 2.
13. Alan M. Rugman, 'Business Concerns About Implementing the Free Trade Agreement', *Business Quarterly*, Spring, 1989, p. 25.
14. Susan Sternthal, 'Mexican Fears of Open Trade are Subsiding', *Washington (D.C.) Times*, 11 June 1990.
15. Diane Lindquist, 'Mexican Leader Pushing Even Harder for Free Trade', *San Diego Union*, 6 September 1990.
16. *Time*, 10 August 1992.
17. Susan Walsh Sanderson, 'Mexico – Opening Ahead of Eastern Europe', *Harvard Business Review*, September–October 1990, p. 33.
18. Khosrow Fatemi and Jim Giermanski, 'The Maquiladora Industry', presented at Association for Global Business, Orlando, FL, November 1990.
19. Jim Carlton, *The Wall Street Journal*, 23 September 1992, p. R-16.
20. Katherine Ellison, 'Eyes on the Future, Mexico Changes Rapidly', *The Charlotte Observer*, 31 March 1991, p. 13A.
21. Mike Davis, 'U.S.–Mexico Seek Free Trade Accord', *Houston Post*, 28 March 28 1990.
22. Katherine Ellison, p. 13A.
23. Kevin Merida, 'U.S., Mexico to Explore Agreement on Free Trade', *Dallas Morning News*, 12 June 1990.

24. Paul Power, 'Farmers Object to Free Trade with Mexico', *Tampa Tribune*, 27 September 1990.
25. Stuart Auerbach, 'Bilateral Trade Packets Worry Experts', *Washington (D.C.) Post*, 3 November 1988.
26. Jeffery J. Schott, *More Free Trade Areas*, (Washington, DC: Institute for International Economics, May 1989) p. 11.
27. Diane Lindquist, op cit.
28. Susan Sternthal, 'Trade Pact Looking Better to Balkers South of the Border' *(Washington, DC) Insight/Washington Times*, 18 June 1990.
29. Karen E. Thuermer, 'Global Trade', vol. III, no. 2, February 1991.

6 North American Trade in the Post-Debt-Crisis Era

Barry W. Poulson and Mohan Penubarti

6.1 INTRODUCTION

Ordinarily we think of financial crises as having a negative impact on trade flows as well as financial flows. Charles Kindleberger, for example, provides the historical evidence of several centuries to show that periods of crisis in international financial markets have been accompanied by the collapse of international trade as well (Kindleberger, 1978). The Great Depression of the 1930s is the classic case of a downward spiral in world trade, tied to the collapse of international financial markets. Based upon historical precedent we would predict that the financial crisis of 1982 would also have had a disruptive impact on world trade. Given the fact that the 1982 crisis was triggered by the Mexican government's inability to service its foreign debt, and that that debt was held mainly in the United States, we would expect the crisis to have been particularly disruptive to Mexican–U.S. trade.

However, historical experience may not prove to be a good guide in predicting the impact of the 1982 debt crisis on North American trade. That crisis was not accompanied by widespread default on international loans by the debtor countries, as had occurred in previous financial crises. The refinancing and rescheduling of the external debt of the major debtor countries established new game rules for international finance. The result has been even greater cooperation between the creditor and debtor nations designed to stabilize and expand international financial markets.

Increased cooperation among the North American economies in their trading relationships has accompanied the rescheduling and restructuring of Mexico's external debt. It is fair to say that the success of the restructuring of Mexico's debt has depended to a great extent upon Mexico's ability to sustain and expand trade with its major trading partner and creditor, the United States. Mexico has been able to continue servicing its external debt from the foreign exchange earnings generated by an expansion in exports, primarily to the United States. That expansion of exports, in turn, has required access to credit in international financial markets that has been forthcoming because Mexico has continued to service its external debt.

The expansion of Mexican trade in the post-debt-crisis era has taken place in an entirely different policy regime compared with that before the crisis. In part because of the need to expand exports, Mexico has abandoned many of the inward-looking policies it had pursued since the Second World War. While there is still a strong dirigiste element in Mexican trade policy, it is fair to describe that policy as one of export promotion, compared with the import substitution policies of the pre-debt-crisis period. Mexico has significantly reduced barriers to trade through bilateral negotiations, as well as through GATT.

Trade between the United States and Canada since the early 1980s has also been marked by increased cooperation. Liberalization of trade between the two countries, of course, preceded the debt crisis in 1982; however, significant reductions in trade barriers have occurred during the last decade. The Canada–U.S. free trade agreement represents the culmination of a series of negotiations between the two countries to reduce barriers to trade and factor flows. The United States, Canada and Mexico are currently negotiating a North American free trade agreement to reduce multilaterally the barriers to trade and factor flows.

The liberalization of North American trade since the early 1980s should lead to patterns of trade that more closely approximate the comparative advantage of each of the North American countries. However, trade liberalization up to this point has been dominated by bilateral negotiations. Thus, recent North American trade presents a unique case study of the impact of bilateral, as opposed to multilateral, trade liberalization. Theoretically, bilateral negotiations to reduce trade barriers should give rise to two effects: *trade creation*, or the displacement of domestic production in the importing country by imports from the other negotiating country; and *trade diversion*, or the displacement of imports from a third country by imports from the negotiating country. The latter effect will be important if and only if a third country exports similar products to the markets of the negotiating countries. If exports are dissimilar then there is less room for commodity overlap and trade diversion.

6.2 ESTIMATION

In this section we report the results of tests of the changing patterns of North American trade before and after the debt crisis. Market

Economy Trade data compiled by the United Nations and published in the *International Trade Statistics Yearbook* were used for the 1962–90 period. The 3-digit SITC export series were aggregated into four broad trade categories comprising (1) primary exports (food and live animals; beverages and tobacco; crude materials, excluding fuels; animal, vegetable oil, fat), (2) mineral fuels, (3) basic manufactures (chemicals; basic manufactured goods), and (4) advanced manufactures (machines, transportation equipment).

The first step involved modelling the export series for each of the four trade categories to examine the patterns for the 1962–90 and 1982–90 periods among the North American countries. The export series was estimated as a linear model whose error term is assumed to be an autoregressive process of a given order p, denoted by $AR(p)$. The general $AR(p)$ model is specified for each time t as:

$$Y_t = X_t'\beta + v_t$$

where

$$v_t = \epsilon_t - \alpha_1 v_{t-1} - \ldots - \alpha_p v_{t-p}$$

and Y_t is the dependent variable, X_t is a vector of regressor values, β is a vector of structural parameters, and ϵ_t is assumed to be normally and independently distributed with mean zero and variance σ^2 (see Harvey, 1981, or Mills, 1990). The annual trade data was modelled in first differences with each of the broad export categories comprising the dependent variable. A dummy separating the pre-debt-crisis period (1962–81) from the post-debt-crisis period (1982–90) was used as a regressor.

The next step in the analysis was to measure changes in the similarity of exports of the North American countries before and after the debt crisis. Changes in measures of export similarity can be used to assess the degree to which there are overlapping trade patterns, and the extent to which their economic structures are becoming more or less divergent. The measure of export similarity used is that suggested by Finger and Kreinen (1979) and is defined by the formula

$$S(ab, c) = \left\{ \sum i \text{ Minimum } [Xi\,(ac), Xi\,(bc)] \right\} 100,$$

which measures the similarity of the export patterns of countries a and b to country c. $Xi(ac)$ is the share of commodity i in country a's exports to country c and, similarly, $Xi(bc)$ is the share of commodity i in country b's exports to country c. If the commodity distribution of exports of country a and country b to country c are identical ($Xi(ac)$ = $Xi(bc)$ for each i), the Finger–Krienen index will take on a value of 100. On the other hand, if the export patterns are totally dissimilar, then the index will take on a value of zero.

The Finger–Krienen index was also estimated as a linear model with an autoregressive error term to examine the changing patterns in export similarity, if any, before and after the debt crisis. The results for these two sets of estimations are presented in the following section.

6.3 TRADE PATTERNS

Tables 6.1, 6.2 and 6.3 present the results from fitting the AR(1) model for the patterns of North American trade. In each case the constant term captures the trend in trade for the period as a whole, while the dummy variable fitted for the period 1982–90 captures the changing pattern of trade after the debt crisis.

6.3.1 U.S. Exports

Table 6.1 shows the evidence for United States exports. The rate of growth of U.S. exports to Mexico shows a positive trend for all commodities for the period as a whole. The period since 1982 shows no significant change in the rate of growth of trade for any commodity. The 1982 debt crisis would appear to represent at most a temporary interruption in the long-term trends in U.S. trade with Mexico. The major impact of bilateral trade negotiations between the U.S. and Mexico since the debt crisis would appear to be in sustaining this long-term trend of growth in U.S. exports to the Mexican market. We must qualify this inference by noting that the data covers the relatively short period since recent bilateral trade negotiations were completed between the two countries.

There is a significant positive trend in U.S. exports to Canada covering the period as a whole. Since 1982, however, the only exports showing an acceleration in the rate of growth are mineral fuels. The lack of evidence for acceleration in exports of manufactured goods over the last decade is surprising in light of the Canada–U.S. FTA. That agreement appears to have merely sustained the growth of U.S. exports to Canada over the period as a whole. Again, we must qualify

Table 6.1   Estimates of U.S. export patterns

Importing country	Export category			
	Primary exports	*Mineral fuels*	*Basic manufactures*	*Advanced manufactures*
Canada:				
Constant	0.12	0.03	0.30	0.54
	(5.38)**	(4.45)**	(18.07)**	(16.47)**
Dummy	0.002	0.01	0.02	0.01
	(0.17)	(2.41)**	(0.79)	(0.16)
$\bar{R}^2$	0.93	0.75	0.49	0.64
D–W	1.86	2.20	1.89	1.80
Mexico:				
Constant	0.15	0.04	0.33	0.49
	(13.41)**	(11.65)**	(29.69)**	(64.36)**
Dummy	0.002	0.003	−0.003	−0.01
	(0.15)	(0.45)	(−0.17)	(−0.97)
$\bar{R}^2$	0.10	0.01	0.41	0.14
D–W	1.86	1.99	1.96	1.92

Note: Numbers in parentheses are t-ratios. ** indicates significance at the 0.05 level and * indicates significance at the 0.10 level.

this inference by noting that the data covers a relatively short time period since that agreement.

6.3.2   Mexican Exports

Table 6.2 shows the results for Mexican exports. For the period as a whole Mexican exports to the United States show a significant positive trend for all categories of goods. There is no evidence of a significant change in the trend of growth of exports of any commodity from Mexico to the United States following the debt crisis in 1982. Bilateral liberalization of trade between Mexico and the United States since 1982 appears to have merely sustained the long-term trend of growth in Mexican exports to the United States. Again we must qualify this inference by noting the short time that has elapsed since the most recent bilateral negotiations between the two countries.

This evidence may be surprising given the the rapid growth in manufactured exports from Mexico to the United States since the early

Table 6.2   Estimates of Mexican export patterns

Importing country	Export category			
	Primary exports	*Mineral fuels*	*Basic manufactures*	*Advanced manufactures*
United States:				
Constant	0.45	0.21	0.21	0.10
	(2.11)**	(1.86)**	(8.197)**	(1.87)**
Dummy	−0.03	0.09	−0.0251	0.06
	(−0.39)	(1.23)	(−0.726)	(1.10)
$\bar{R}^2$	0.90	0.92	0.72	0.70
D−W	2.36	2.08	1.93	2.17
Canada:				
Constant	0.37	0.08	0.36	0.18
	(8.57)**	(4.0)**	(12.30)**	(3.39)**
Dummy	0.24	0.11	−0.21	−0.14
	(2.80)**	(2.28)**	(−3.51)**	(−1.45)
$\bar{R}^2$	0.25	0.50	0.38	0.36
D−W	1.63	1.74	2.05	1.92

Note: Numbers in parentheses are *t*-ratios. ** indicates significance at the 0.05 level and * indicates significance at the 0.10 level.

1980s. But we must recall that Mexico experienced rapid growth in exports of manufactured goods to the United States prior to the debt crisis. In the years leading up to the oil boom and debt crisis Mexico was viewed as one of the new exporting developing countries because of its success in expanding exports of manufactured goods. The oil boom and the debt crisis represent at most a temporary disruption in this long-term upward trend in the growth of manufactured exports from Mexico to the United States.

In recent years Mexico has been referred to as the young tiger of the Western Hemisphere. Mexico appears to be following the path of the Asian NICs in rapidly expanding its exports of advanced manufactured goods to the U.S. market. Advanced manufactures now account for 31 per cent of Mexico's total exports, which exceeds the share of exports accounted for by fuels. Many of these goods are component parts produced in maquiladora firms, and much of this is intra-industry trade, for example in the automotive industry. These industries use abundant supplies of low-cost labor, as our theories of comparative advantage predict. However, Mexico has recently demonstrated the capacity to rapidly expand its exports of semi-assembled and final products produced by a number of advanced manufacturing industries as well. With the rapid growth in the export of manufactured goods to the U.S. market the Mexican economy has become more closely integrated with that of the United States.

Mexican exports to the Canadian market show a quite different pattern. For the period as a whole there is a significant positive trend in the growth of exports of all commodities. Since 1982 the evidence shows a significant increase in the rate of growth of primary goods and fuels, and a significant decrease in the rate of growth of exports of basic manufactures. The volume of Mexican trade with Canada has always been small relative to trade with the United States, which may account for the greater volatility of that trade. Nonetheless the implication is that the debt crisis has been followed by discontinuous shifts in the trend of Mexican trade with Canada, in contrast with the continuity in its trade with the United States.

Several factors have contributed to the discontinuous changes in the rate of growth of Mexican exports to Canada since 1982. To some extent these changes reflect an attempt on the part of the Mexican government to diversify its foreign markets. This would account for the discontinuous increase in the rate of growth of mineral fuels, and possibly primary goods, exported to Canada since 1982. The significant decline in the rate of growth of exports of basic manufactures is surprising. The Canadian market accounts for a small share of exported basic manufactured goods, and an even smaller share of advanced manufactured exports from Mexico. This may account for the volatility in those exports over time. However, the evidence suggests that the debt crisis did significantly disrupt the upward trend in manufactured exports to Canada, and Mexico has yet to recover from that disruption. It is surprising that Mexico has not been successful in capturing a larger share of the Canadian market for basic manufactured goods, particularly in light of their success in penetrating the U.S. market for manufactures. One possible explanation is that Mexico and Canada have increasingly duplicated each others' exports of basic manufactured goods since 1982. We will test this hypothesis more rigorously later in this chapter. Certainly one explanation is the fact that Canada and Mexico have not entered into the kind of bilateral negotiations to liberalize trade in manufactures that have been undertaken between Mexico and the United States.

6.3.3 Canadian Exports

Table 6.3 shows the results for Canadian exports. For the period as a whole Canadian exports to the U.S. market show a significant positive trend for all categories of goods. Since 1982 the only significant change in these trends is a decline in the rate of growth of exports of basic manufactured goods. The continuity in the trend of growth for most Canadian exports to the United States is consistent with the evidence of continuity in U.S. exports to the Canadian market.

Table 6.3   Estimates of Canadian export patterns

Importing country	Export category			
	Primary exports	*Mineral fuels*	*Basic manufactures*	*Advanced manufactures*
United States:				
Constant	0.25	0.12	0.33	0.34
	(2.23)**	(5.49)**	(8.41)**	(5.65)**
Dummy	−0.01	0.01	−0.003	0.03
	(−0.45)	(0.41)	(−2.10)**	(1.03)
$\bar{R}^2$	0.96	0.74	0.89	0.93
D−W	2.08	1.95	2.19	2.22
Mexico:				
Constant	0.35	0.19	0.35	0.06
	(9.62)**	(2.39)**	(11.10)**	(1.63)**
Dummy	0.09	−0.02	−0.14	0.12
	(1.42)	(−0.20)	(−2.49)**	(2.04)
$\bar{R}^2$	0.29	0.40	0.34	0.45
D−W	1.80	2.16	2.18	2.08

Note: Numbers in parentheses are t-ratios. ** indicates significance at the 0.05 level and * indicates significance at the 0.10 level.

Canada and the United States have become more closely integrated as a result of trade liberalization over the last decade, but the evidence suggests that Canada is becoming less specialized in basic manufactures. This result may appear to be counterintuitive given prior expectations of trade in manufactured products between the two countries. However, a substantial body of literature predicted that

Canadian producers of basic manufactured goods would be at a competitive disadvantage under a free trade regime with the United States. These basic manufactured goods are subject to economies of scale and long production runs where Canadian firms are at a disadvantage vis-à-vis U.S. firms. On the other hand the stability in the growth of trade in advanced manufactures between the two countries reflects the growing importance of intraindustry trade. On the whole the evidence suggests complementarity in trade between Canada and the United States. Again, we will test this hypothesis more rigorously at a later point in this chapter.

Canada's exports to the Mexican market for the period as a whole show a significant positive trend for all commodities. Since 1982, however, the evidence suggests some discontinuous changes in the trend of exports in manufactured goods. The rate of growth of exports of basic manufactures has fallen, while that for advanced manufactures has increased. This evidence suggests that Canadian producers of basic manufactures have been less competitive in the Mexican market, as well as in the U.S. market, in recent years. On the other hand, Canadian producers of advanced manufactured goods have been more successful in the Mexican market since 1982. The latter success reflects the expanded role of the Maquiladora industry and intra-industry trade in advanced manufactures between Canada and Mexico.

Canada did little towards negotiating a liberalized trade regime with Mexico over this period, so that the patterns of trade between the two countries may be more distorted by government intervention. The evidence suggests that Canadian exports of manufactured goods may increasingly duplicate rather than complement that of Mexico. We will test this hypothesis in the following section.

6.3.4 Similarity in Trade Patterns

Table 6.4 summarizes the results of the AR(1) model fitted to measures of similarity in North American trade. Figure 6.1 traces the index of similarity of exports for the three North American countries. Figures 6.2–6.7 in the appendix provide evidence for the source of these trends in the similarity index, tracing the share of each country's total exports accounted for by each of the major commodity groups. These measures bear out some of the conjectures suggested in the above analysis regarding trends in the similarity of trade between these countries.

The index of similarity in U.S. and Canadian exports to Mexico is relatively high (54 per cent) and stable over the entire period. This is

Table 6.4 Similarity in trade patterns

	Can./Mex.→U.S.	Can./Mex.→Mex.	U.S./Mex.→Can.
Constant	62.11	53.78	62.42
	(15.32)**	(17.29)**	(13.12)**
Dummy	5.69	5.72	−21.19
	(0.89)	(1.00)	(−2.41)**
$\bar{R}^2$	0.38	0.13	0.16
D−W	1.97	2.00	2.10

Note: Numbers in parentheses are t-ratios. ** indicates significance at the 0.05 level and * indicates significance at the 0.10 level.

Figure 6.1 Finger–Kreinen index for export patterns

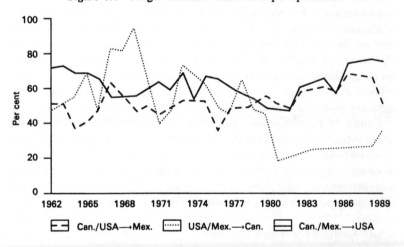

what we would expect in the exports from these two industrialized countries to an LDC such as Mexico.

What is unexpected is the high index of similarity in Canadian and Mexican exports to the U.S. market. This index, at 62 per cent, is even higher than that for Canadian and U.S. exports to Mexico. This evidence supports the hypothesis introduced earlier, that Canada and Mexico increasingly duplicate each others exports' to the United States. This might explain the reluctance of Canada and Mexico to negotiate bilateral and multilateral agreements for trade liberalization comparable to the Canada–U.S. FTA.

The index of similarity in U.S. and Mexican exports to Canada also is high at 62 per cent, but it has declined significantly since 1982. The implication is that recent industrialization has resulted in a growing complementarity rather than duplication in the trade of these two countries. This might explain the greater willingness of the United States and Mexico to enter into both bilateral and multilateral agreements to liberalize trade. Trade liberalization, in turn, may have contributed to complementarity in trade consistent with comparative advantage in the two countries.

6.4 CONCLUSIONS

The historical experience regarding the effects of financial crises on trade does not provide much of a guide in predicting the impact of the 1982 debt crisis on patterns of North American trade. The United States has maintained a high rate of growth in exports of all commodities to her North American partners, and there is no evidence of a significant change in these trends in trade following the debt crisis. Mexican exports to the U.S. market also show high rates of growth, with no evidence of a significant shift following the crisis.

Trade between Canada and Mexico, on the other hand, shows greater discontinuity in the period following the debt crisis. That trade is small relative to these countries' trade with the United States, which may account for the greater volatility of that trade over time. However the indexes of similarity suggest increasing competition between Mexican and Canadian industries for a greater share of the North American market. This finding helps to explain the reluctance with which Canada seems to have approached the negotiation of the North American Free Trade Agreement.

In contrast, the evidence suggests declining similarities in U.S. and Mexican exports to Canada since the debt crisis. The implication is that expansion of the Canadian market is accompanied by trade creation for both the United States and Mexico, since there is less overlap in their exports to that market. Both the United States and Mexico would appear to have an incentive to negotiate the North American Free Trade Agreement, not only to expand trade between themselves, but also to expand their trade with Canada.

6.5 APPENDIX

Figure 6.2 Mexican exports to the United States (1962–90)

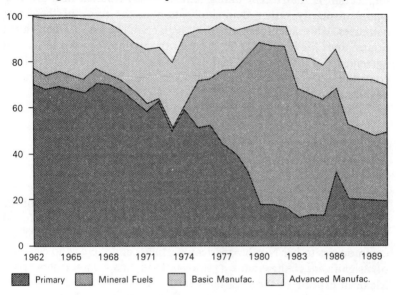

Figure 6.3 Mexican exports to Canada (1962–90)

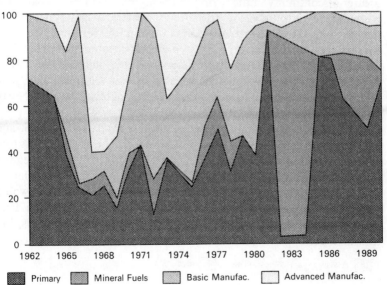

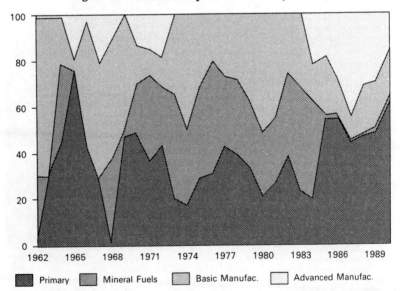

Figure 6.4   Canadian exports to the United States (1962–90)

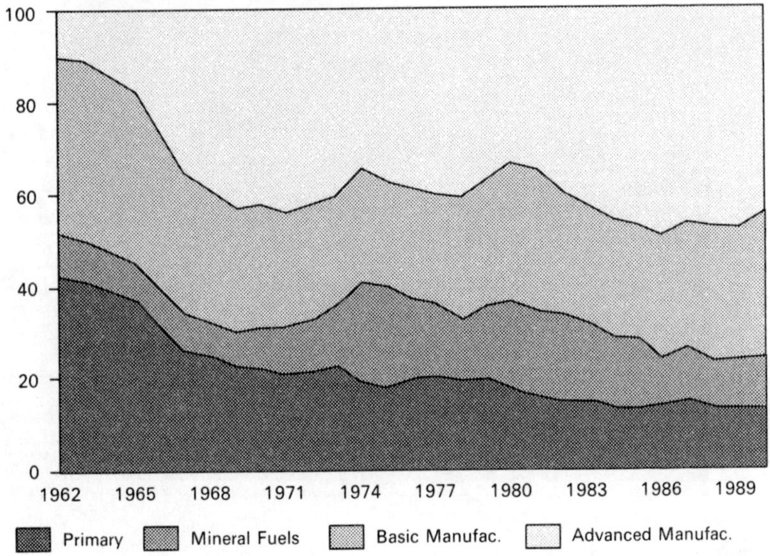

Figure 6.5   Canadian exports to Mexico (1962–90)

Figure 6.6   U.S. exports to Canada (1962–90)

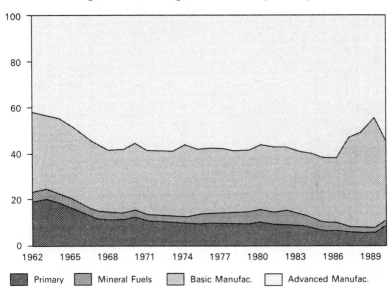

Figure 6.7   U.S. exports to Mexico (1962–90)

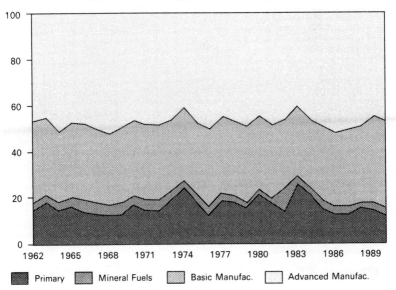

Part III

National Perspectives and Bilateral Issues

7 Introduction

The benefits and/or costs of NAFTA notwithstanding, the three member-states of the agreement are independent nations with individual, and sometimes conflicting, perspectives. Furthermore, their partnership in NAFTA notwithstanding, the bilateral issues between the countries remain strong and in force. Part III of this volume examines these specific issues. In Chapter 8, Ricardo Grinspun analyzes the impact of the Canada–U.S. Free Trade Agreement – and potentially of NAFTA – on the Canadian manufacturing sector. He maintains that some provisions of the Canada–U.S. FTA will not be fully implemented until the end of the decade, and therefore it is too early to pass final judgement on the effects of the agreement on the Canadian economy. Nevertheless, Grinspun uses the experience of the first four years of the agreement to reach certain conclusions. He argues that the Canada–U.S. FTA cannot be evaluated independently from the general economic policies of the Canadian government and that 'a better way to picture the Canada–U.S. FTA is as one element in the overall economic policy of the government.' Using this approach, Grinspun concludes that 'the Canadian government has done it all wrong', and that 'the Canada–U.S. FTA has both costs and benefits; however the economic policies of the government have created a situation where Canada is paying the costs of the agreement but is unable to realize the gains to be had from it'.

In Chapter 9, Edward R. Bruning provides the Canadian perspective of the North American Free Trade Agreement. He identifies Canada's major objectives in the NAFTA negotiations as gaining access to the Mexican market, improved security of access to the U.S. market, improved conditions under which Canadian businesses can make strategic alliances within North America, ensuring Canada's position as an attractive site for foreign investment, and the establishment of a fair and expeditious dispute settlement mechanism. In addition, Bruning identifies several less significant goals that Canada had in negotiating, and will have in implementing, the NAFTA agreement. However, Bruning maintains that these are not universally accepted goals in Canada and the opposition parties, particularly the National Democratic Party and the Liberal Party, have not been supportive of the agreement. Should either party win the next parliamentary election in Canada, the future of NAFTA will be severely jeopardized. The Liberal Party has, in fact, made rescinding the Canada–U.S. FTA a campaign issue.

Chapter 10 contains Joseph A. McKinney's analysis of Mexico's entry into NAFTA. He traces the recent changes in Mexico's economic and commercial policy from the days of import substitution to its entry into GATT in 1986, the signing of the 'Framework of Principles and Procedures for Consultation Regarding Trade and Investment Relations with the United States', and finally the decision to enter into negotiations for the formation of NAFTA. He concludes that 'incorporating Mexico into a North American Free Trade Area has the potential for benefiting not only Mexico, but the United States and Canada as well. [However,] because of the small size of Mexico's economy relative to those of the United States and Canada, the economic impact of a free trade agreement will be much greater for Mexico than for either of the other two countries . . . [But even in the case of the latter two,] these benefits will not be inconsequential.' Among the specific benefits discussed by McKinney are the improved international competitiveness of U.S. and Canadian industries as a result of increased production-sharing using lower Mexican labor costs, the expanded Mexican market brought about by the modernization – and the greater prosperity – that should accompany the agreement, and finally, Mexico's potential to 'serve as a catalyst for closer economic integration and political cooperation throughout the Western Hemisphere'.

In Chapter 11, Peter Morici provides a different view of Mexico's entry into NAFTA. According to Morici, membership in NAFTA manifests different policies and policy objectives by the participating countries in the agreement. Among the three countries, 'Canada has joined the talks, largely to guard against the erosion of its 1989 pact with the United States [the Canada–U.S FTA] . . . [while] for Mexico, free trade would mark the climax of a radical change in development strategy . . . [and] for the United States, a free-trade agreement has the potential to be a key element in a new national policy to foster competitiveness.' After tracing the evolution of economic environment in the three countries, particularly Mexico, Morici provides a discussion of opportunities and adjustments for the United States and Mexico and concludes that 'for decades, Americans have been nagging and cajoling Latin Americans to open up their markets to U.S. goods and investment and let the compelling energies of market capitalism and entrepreneurship transform their societies'. Now that an opportunity has been created by Mexico, it must been taken advantage of. 'The United States must recognize both the opportunities that reform in Mexico offers and that a failure to embrace free trade would adversely affect economic and political progress there. The status quo is not an option, because events are moving so quickly and recovery is so fragile in Mexico that there really is no status quo in Mexico or in U.S.– Mexico relations'.

Jorge G. Gonzalez and Alejandro Velez provide an empirical discussion of intra-industry trade between the United States and Mexico in Chapter 12. Specifically, the authors present an evaluation of the level of intra-industry trade between the two countries. Their indices of intra-industry trade indicates a rapid increase in this type of trade during the 1980s. Additionally, Gonzalez and Velez' study finds that, in comparison with similar indices for other countries, the level of intra-industry between the United States and Mexico is quite high. Gonzalez and Velez conclude that because of the high level of intra-industry trade, any NAFTA-induced expansion of trade should not create any major dislocation of productive activities in the two countries.

Chapter 13 examines an important aspect of bilateral relations between the United States and Mexico, specifically, the question of foreign exchange relations between the two countries. In this chapter, Kurt R. Jesswein, Stephen B. Salter, and L. Murphy Smith compare the use and perceptions of foreign-exchange risk-management techniques among large U.S. and Mexican companies. Their study reveals some significant differences in the utilization of products, perspectives on key factors in product selection and attitudes to innovation. They also explore some of the factors that may influence these differences. Jesswein, Salter and Smith using a questionnaire survey of the 100 largest Mexican companies listed in Duns Latin American Guide and comparing the survey with a similar study of U.S. firms, compared the approach of U.S. and Mexican firms in managing their foreign exchange risk. They concluded that 'in moving south, U.S. companies will find a market that is as or more committed to FERM [foreign exchange risk management] than they are. This market will be familiar, having many of the same products that the U.S. manager uses. [On the other hand,] Mexican managers expanding northward may have to reevaluate their repertoire of FERM techniques, but may be pleasantly surprised, given the relative inactivity of U.S. managers, at how well their previous experience has prepared them.'

8 Free Trade Restructuring in Canadian Manufacturing: An Initial Assessment*

Ricardo Grinspun

8.1 INTRODUCTION

The Canadian government's decision in 1985 to embark on discussions leading to a bilateral free trade arrangement with the United States followed years of debate in academic and policy-making circles on the potential merits and disadvantages of this path (Macdonald Report, 1985; Whalley with Hill, 1985; Doern and Tomlin, 1991). Influential analysis showed that a key sector that would benefit from further trade liberalization was manufacturing. Manufacturing in Canada, as in other places, is characterized by a high market concentration and has the potential for increasing returns to scale. The view was that Canadian manufacturing is suffering from low productivity and poor technological performance due to limitations of the Canadian market and the inability to exploit the returns to scale. Free trade with the United States would provide major benefits to Canadian manufacturing in terms of rationalization of production, lower manufacturing costs, secure access to the main export market, increased international competitiveness, efficiency gains, domestic deregulation and more emphasis on market signals in the allocation of resources (Wonnacott, 1985).

The qualitative analysis of gains from free trade was accompanied by quantification efforts. A study that received major public exposure was carried out by Richard Harris with the assistance of David Cox (1983) – (hereafter Harris–Cox). Their work projected significant economic benefits for Canada from a bilateral free trade agreement with the

---

* The author wishes to acknowledge useful discussions with Fred Lazar, as well as comments by Michael Copeland, Robert Kreklewich, Louis Lefeber and Keith MacKinnon on an earlier draft. The usual disclaimer applies.

United States. Their projection was for an impressive 9.0 per cent gain in aggregate Canadian welfare, a 28 per cent raise in real wages, an almost 30 per cent increase in labor productivity and a 100 per cent increase in the volume of bilateral Canada–U.S. trade. Furthermore, they predicted an increase of 225 per cent in the average output per firm in the manufacturing sector (Harris, 1985b, p. 173). The quantitative work by Harris–Cox and others was influential during the Canada–U.S. FTA negotiation and ratification, since it projected large economic benefits for Canada from such an agreement.[1]

The Harris–Cox study elicited a number of criticisms that pointed to a likely overstatement of gains for Canada from bilateral free trade (Whalley, 1984; Stern, 1985). Work carried out by the Canadian Ministry of Finance using an updated version of the Harris model showed more modest gains for Canada (Government of Canada, 1988). There were also opponents who claimed that the bilateral free trade strategy would have significant costs for Canada (Watkins, 1985; Wilkinson, 1987). They argued that quantitative models tended to underestimate, or avoid, the likely costs of such an agreement. These drawbacks could include income redistribution, painful restructuring, adjustment and relocation, as well as likely political, environmental and social costs. Opponents argued that their cost-benefit analysis showed that bilateral free trade was economically and socially undesirable for Canada.

The purpose of this chapter is to throw some light on this debate, based on the experience of almost four years of implementation of the Canada–U.S. FTA in Canada.* The chapter will proceed as follows. Section 8.2 raises questions about the extent to which Canada obtained access to the huge U.S. market in the Canada–U.S. FTA negotiations. Section 8.3 contrasts the optimistic predictions for productivity gains in manufacturing with disappointing developments during the early 1990s. Section 8.4 focuses on the painful adjustment that the manufacturing sector has undergone, and Section 8.5 considers how a poor macroeconomic environment has distorted that adjustment. Section 8.6 concludes that the benefits to Canada from the Canada–U.S. FTA have been, and are likely to continue to be, very small. The initial evidence suggests the projections by Harris–Cox and others seemed to grossly overstate the gains to be had from the agreement. In

---

* 'Reasonable' macroeconomic policies would strive for low real interest rates and purchasing power parity between the Canadian and US dollars.

contrast, the economic and social costs of free trade have been large, raising serious doubts about the prospects for Canadian economic and social development during the 1990s.

8.2 HOW MUCH TRADE LIBERALIZATION?

The work by Harris–Cox, as well as other early studies completed before the Canada–U.S. FTA was actually negotiated, assumed a policy experiment that completely eliminated barriers to trade between Canada and the United States (that is, true bilateral free trade was achieved). The actual agreement, ratified in 1988, has only had a modest trade-liberalizing effect since it will gradually eliminate tariffs but will maintain many non-tariff barriers (NTBs). Moreover, except for a few protected sectors, tariffs were not a major obstacle to bilateral trade prior to the agreement. The average tariff paid by Canadian exporters to the United States in 1987 was 4 per cent, and about 70 per cent of the exports entered duty free. Out of 19 manufacturing sectors, only one (apparel) confronted U.S. tariffs higher than 10 per cent in that same year.* For this reason, the macroeconomic benefits for Canada from the elimination of U.S. tariffs had to be small.

A major objective of the Canada–U.S. FTA from the Canadian perspective was to protect its exporters from what was perceived as a growing protectionist tendency in the United States, that expressed itself mainly through the application of different trade remedy actions. These included safeguard measures, antidumping and countervailing duties, and 'section 301' actions.[2] The Canadians obtained limited gains in this key area of negotiation.[3] Many U.S. policies, perceived as protectionist in Canada, were unaffected by the implementation of the agreement. The agreement left some policies unchanged and modified others with some improvement for Canada. Emergency measures (or 'fair trade' laws) were partially maintained. No exemption from GATT safeguard measures was obtained (Richard and Dearden, 1987).

A major drawback for Canada was that measures against unfair trade practices based on existing U.S. laws, like anti-dumping and countervailing duties, were legitimized by the agreement. This was an area where Canadian exporters to the United States had suffered

---

* The implementation of the Canada–U.S. FTA started in January 1989, and tariff elimination will be completed in 10 years.

frequent harassment in the past due to long and expensive litigation. The original Canadian objective was to achieve a common definition of unfair pricing and allowable government subsidies. This definition would protect Canada's ability to implement domestic policies, such as regional development programs, which had repeatedly been challenged in the past by the United States as unfair trade practices (Doern and Tomlin, 1991, p. 29).

The two parties could not agree on subsidies and unfair pricing codes, and an interim face-saving approach was quickly put into place in the last hours of the negotiations (Chapter 19 of the Canada–U.S. FTA). Under this approach, each country continued to apply its own antidumping and countervailing duty laws, but a binational dispute settlement mechanism was created to ensure that the actions of each country were consistent *with its own laws*. This was one of several instances where the Canada–U.S. FTA legitimized the use of U.S. trade laws toward Canadian exporters, rather than exempting them. A binational working group was formed to try to find a permanent solution to the issues, including the thorny one of subsidies to trade, but these negotiations never got anywhere. The NAFTA negotiations completed August 1992 enshrined the Canada–U.S. FTA dispute settlement mechanism as a permanent feature of the new agreement, which also includes Mexico.[4]

Those Canadians who supported this arrangement hoped that the application of contingency protection in the United States would be depoliticized, creating a speedier, more effective resolution of disputes. The intensive use of Chapter 19 panels since the implementation of the agreement shows some gains for Canada in terms of speed and effectiveness. But it also shows that the 'harassment' of Canadian exporters also continues, as the recent pork case demonstrated.* The binational panel ruled in favor of Canada and an 'extraordinary challenge' of the panel's results by the United States failed to change the outcome.

Although the challenge committee upheld the panel's remand, this was a clear warning to Canadian pork producers that increased exports into the U.S. would be met by further trade actions, under one or more trade laws (Sinclair, 1993).

---

* The Canadian tariffs were somewhat higher. U.S. exporters to Canada faced an average tariff of about 5 per cent. Source for data is Hart (1990, pp. 42–3).

The likelihood of future trade actions raises uncertainty about future market access and dampens plans for expanded exports into the U.S. market. The problem for the Canadian exporters is the legitimacy and intensity of application of U.S. contingency protection – and the agreement seems to have worsened things for Canada on both counts.[5]

Moreover, the agreement certainly did not spare Canada from protectionist pressures in Congress. Even in the midst of approval of the agreement, Congress passed the 1988 Omnibus Trade and Competitiveness Act and did not exclude Canada from the effects of this protectionist bill, although some limited exemption was achieved in terms of future legislation. Recent controversial interpretations of the rules of origin in the automotive sector have permitted the United States to impose restrictions on the exportation to the United States of Japanese cars assembled in Canada.

In summary, the major trade-liberalizing effect of the Canada–U.S. FTA was the gradual elimination of tariffs – which were already very low. In contrast, liberalization in the area of contingency protection and NTBs was disappointingly limited. Even ardent supporters agreed that the agreement did not achieve secure access to the U.S. market; the debate in Canada focuses on whether or not the degree of improvement achieved is meaningful.

8.3 PRODUCTIVITY

Productivity improvements are considered to be a crucial benefit from freer trade. Wonnacott (1985, p. 69) has described how free trade with the United States would create these productivity gains in a context of a Canadian manufacturing sector that operates under increasing returns to scale:

> With the elimination of trade barriers between the two countries, Canadian producers would rationalize. That is, they would specialize in a smaller range of goods, producing each at higher volume and frequently at lower cost. In doing so, Canadian producers would be responding to both a carrot and a stick. The carrot would be free access to the large U.S. market. The stick would be the removal of the Canadian tariff, which would leave many Canadian firms unable to compete with less expensive imports unless they did rationalize and thus reduce their costs.

Even if one accepts the assumptions upon which these predictions are based, the quantitative significance of the productivity gains is of importance. How much improvement should we expect from a particular trade-liberalizing scenario?

Early trade policy experiments predicted significant increments in Canadian productivity under various free trade scenarios: unilateral Canadian free trade; bilateral free trade with the United States; and multilateral free trade under GATT (Hill and Whalley, 1985, pp. 24–36). As noted, Harris–Cox predicted a remarkable 30 per cent increment in Canadian labor productivity under bilateral free trade with the United States (Harris, 1985b, p. 173). Critics responded with various reasons for a likely overstatement of productivity gains in the Harris–Cox projections. There were several technical aspects of their computational free trade experiment that unrealistically magnified the gains. These included their assumptions on pricing behavior and parameters representing economies of scale, as well as on the initial level of barriers to trade (Whalley, 1984; Stern, 1985; Wilkinson, 1987).

Canadian economic models tend to focus on trade barriers as a key cause of productivity differentials, and in so doing they abstract from other differences between the productive environments of Canada and the United States. For example, they may assume that technology and production functions are similar in the two countries, the differences in productivity arising mainly from one country producing at a lower rate because of the limitations of the domestic market (Baldwin and Gorecki, 1985, p. 189). Consequently, in a situation of perfect capital mobility and no barriers to trade, these models would predict the convergence over time of productivity levels everywhere in the free trade area. In the case of a small economy being integrated with a large one, the implication is productivity in the small economy will converge with that in the large economy. This analysis abstracts from institutional, technological, human capital and labor market characteristics, and other such elements that create differences in production costs even when there are no barriers to trade. A clear counter-example is that regional differences in productivity within the United States remain, even after a long history of relative free trade and factor mobility.

One way to test the hypothesis that elimination of bilateral trade barriers causes a convergence of productivity levels on both sides of the border is to observe the Canadian experience prior to the Canada–U.S. FTA. The history of Canadian trade after the Second World War is one of continued liberalization promoted through the multilateral trade negotiations under GATT. In particular, trade between the

United States and Canada liberalized significantly in the postwar era and continued to do so during the 1970s and 1980s (Bordé and Cross, 1989). If this 'convergence hypothesis' on productivity were correct, then one would be able to observe an ongoing closing of the productivity gap between Canada and the United States.[6] The evidence on productivity trends, however, is mixed. Using data from the 1970s, Baldwin and Gorecki (1985, p. 190) report that

higher import penetration is associated with greater Canadian relative efficiency. Hence the openness of the Canadian economy positively affects Canada's relative efficiency [vis-à-vis the U.S.].

However, during the 1980s the gap in labor productivity between the two countries did not continue to close, but widened, even though the liberalization trend continued.[7] In terms of total factor productivity (TFP), the gap closed as Canada caught up between 1974 and 1979, but it again widened between 1980 and 1988 (Rao and Lemprière, 1992b, p. 19). In terms of unit labor costs, Canada had a disadvantage of about 26 per cent relative to the United States in 1990 (Rao and Lemprière, 1992a, p. 42).* This productivity gap remains even though the Canadian economy is extremely open and there is a high degree of free trade with the United States. In short, the 1980s provide no empirical support for the convergence hypothesis. This means that a key theoretical ground for the expected gains for Canada from a free trade agreement is weak.

Canada is currently in the fourth year of implementation of the Canada–U.S. FTA and it is useful to relate to recent productivity developments. Labor productivity in Canada has continued its upward trend since free trade was implemented in 1989. However, this upward movement is due mainly to a significant increase in the capital/labor ratio in production (Statistics Canada, 1992, p. 8). As mentioned before, the labor productivity gap with the United States continued to widen during this period. The increase in the capital/labor ratio is principally a response to the recessionary conditions in Canadian manufacturing since 1989, which have expressed themselves in rising unemployment and lower capacity utilization rates. The recent increase in the capital/labor ratio is *not* the result of accelerated investment in

---

* Comparisons of unit labor costs are affected by exchange rate changes, but Rao and Lamprière estimate that a significant part of the gap in unit labor costs is due to productivity differences.

new plant and equipment, since recent aggregate rates of investment have been quite disappointing.*

The most important variable to observe is TFP, which shows the increment in production that cannot be explained by increases in factors of production. If free trade restructuring in Canada were proceeding toward a more efficient, competitive manufacturing sector, TFP would be the key variable to take an upturn over trend values. The expected benefits from free trade in terms of rationalization of production and increased production scales should inescapably materialize as increments in TFP. In contrast, the recent evolution of TFP is highly disappointing, since there has not only been a reversal of the upward trend of the middle 1980s, but also an absolute fall in the TFP index in each of the years 1989–91 (Statistics Canada, 1992, pp. 8, 87). This is the clearest – although not the only – indicator that the restructuring of manufacturing in Canada was *not* proceeding according to the blueprint sketched by economists in their work prior to the FTA.

In summary, there is no doubt that productivity is affected by flows of goods and factors of production. However, the relationship is not linear and there are other explanatory variables for the level of productivity in a particular country beyond the economy's degree of openness (Denny and Fuss, 1982). It is entirely possible to have ongoing liberalization of trade and the maintenance, or even worsening, of productivity gaps, as happened in Canada during the 1980s. It is highly likely that the recent downturn in TFP was driven mainly by macroeconomic events, which will be discussed in a later section. However, the implication will be that Canada is not obtaining the benefits expected from free trade.

8.4 COSTS OF ADJUSTMENT AND ADJUSTMENT POLICIES

One of the problems with trade liberalization is the inevitable restructuring of production and its implied costs of adjustment, in particular for workers and their families. Harris–Cox projected that 6 per cent of the Canadian labor force would need to move between sectors of production in their multilateral free trade simulation (Harris,

---

* Business investment as a percentage of GDP has shown a definitive downward trend since the third quarter of 1988 to the latest reported quarter – 1992:2 (Statistics Canada, *Canadian Economic Observer*, September 1992, p. 6.5).

1985b, p. 173). Harris (1985a) recommended the implementation of adjustment policies such as new, stronger job retraining and relocation programs. Beyond the expected intersectoral adjustments, free trade would also require significant intrasectoral adjustments due to rationalization of production. In particular, many workers would move from one line of production to another, or from one firm to another in the same sector, as the manufacturing sector specializes in a smaller range of products produced at larger scale. An unfavorable aspect of the Canada–U.S. FTA (as well as NAFTA) from Canada's standpoint is that it does not exempt targeted adjustment policies from challenge under U.S. trade laws. As mentioned, the question of subsidies remains unresolved, and Canadian policies that favor or facilitate adjustment may be interpreted as subsidies to exports under U.S. trade laws. For example, if the Canadian government establishes training programs for laid-off textile workers to facilitate their move to an export-oriented industry, it is possible that this program would be challenged by the United States as being an unfair trade measure.

The fact remains the Canadian government has not implemented special adjustment programs for the Canada–U.S. FTA (which had been promised prior to implementation), arguing that Canada already had effective programs in place. Whether the possibility of a U.S. challenge was a factor in the Canadian government's decision not to implement new adjustment programs is unclear. However, in the absence of strong programs, the social and economic burden of adjustment falls on sectors and people who are likely to be the least prepared to handle them. Sectors characterized by large, relatively unskilled employment are likely to suffer the most disruptive effects of trade liberalization. Labor in these sectors does not transfer easily to modern and efficient sectors, and without government intervention the social cost of weakening these sectors can be significant.

8.5 THE IMPACT OF THE MACROECONOMIC ENVIRONMENT

International trade theory is often presented in isolation from the macroeconomic environment. In the real world of trade policy, however, one cannot exclude the macroeconomic context and policies. It is certainly puzzling that documents like the Canada–U.S. FTA or NAFTA, with their breadth and long-range implications for the countries involved, abstract completely from macroeconomic issues. The fact remains that international trade flows can be extremely sensitive to macroeconomic variables, and in particular to real exchange rate behavior. Given the small trade-liberalizing effects of the agreement, it is clear that macroeconomic developments were likely to swamp the short- and medium-term effects of the new trading regime. Recent Canadian macroeconomic experience provides ample evidence for this belief. The combination of restrictive monetary policy and a large fiscal deficit during the late 1980s brought an increase in short-term and long-term nominal (as well as real) interest rates. One consequence was an increment in the interest rate differential between Canada and the United States. This differential grew considerably during 1989 and 1990, in particular, for short-term interest rates (the trend reversed, though, in 1992). The interest rate differential as well as large inflows of financial capital (promoted in part by the Canada–U.S. FTA) contributed to the Canadian dollar's appreciation vis-à-vis the U.S. dollar. The Canadian dollar went up from an average U.S.$0.72 in 1986 – when the agreement was being negotiated – to U.S.$0.83 in January 1989, to U.S.$0.86 in the second quarter of 1990, to a high of U.S.$0.89 in 1991. During 1992 there was a reversal of this trend, the Canadian dollar falling to U.$.0.80 in October.

These policies certainly worsened the recession in Canada. The real appreciation of the Canadian dollar neutralized any increased competitiveness enjoyed by Canadian exporters following tariff liberalization. Unit labor costs in Canadian manufacturing, measured in U.S. dollars, reached their highest level in years in 1990 (Rao and Lemprière, 1992a, p. 5). The recessionary environment also harmed the restructuring expected from the new trading agreement. Highly protected and inefficient sectors were less resilient since they suffered from the joint pressures of the recession – dollar appreciation, high real interest rates and falling domestic protection. The outcome was plant closures and a large pool of unemployed labor (the official Canadian unemployment rate went up from 7.5 per cent in 1989 to about 12 per cent in 1992). Many of these plants were moved to the United States and some to the maquiladora sector in Mexico.* Appliances, textiles, furniture automobile parts are examples of losing sectors (Campbell, 1993).[8]

---

* The Canadian Labour Congress keeps a partial record of Canadian manufacturing plants that have closed, and in many cases can identify new plants in the United States, Mexico and elsewhere that have replaced them.

In contrast, investment in new plant and equipment – as well as the technological improvements needed for plant restructuring and exploitation of economies of scale – was seemingly not happening under the difficult conditions of the early 1990s. These conditions were not favorable to the investment in, and growth of, new export-oriented production lines that would absorb workers freed from inefficient plants. In the best-case scenario, the hope was that this investment was merely being postponed until better times. In the worst and more likely case, much new investment was being channelled to other places, since Canada had lost much of its attractiveness as an export platform to the United States.

This combination of forces precluded the appropriation of gains from freer trade in Canadian manufacturing, and also pushed this sector into its worst recession since the Great Depression. The statistics are certainly telling: there was a loss of about 26 per cent of total manufacturing employment between June 1989 and March 1992 (Campbell, 1993). According to the government of Ontario, most job losses in that province were due to plant closures and not to lay-offs.

In summary, developments in the Canadian manufacturing sector during the first few years of the Canada–U.S. FTA implementation diverged dramatically from the predicted surge in investment activity and jobs. The picture was, instead, one of a major disruption of this sector, severe loss of manufacturing employment, and relocation of plants to other countries. Macroeconomic distortions were the main cause of this state of affairs. The expected gains from freer trade in manufacturing have not been realized in this environment, and at least some of them may be gone forever (for example, investment flows that have been diverted to other countries).

8.6 CONCLUSION: AN INITIAL EVALUATION OF THE CANADA–U.S. FTA

The implementation of the Canada–U.S. FTA is involving long-term restructuring of the Canadian economy. Some of the agreement's provisions, such as tariff reductions, will be completed by the end of the decade. Consequently many years will pass before a final evaluation will be feasible. However, almost four years into its implementation, one can make an initial evaluation from Canada's viewpoint.

It is easy to dismiss some of the views held about the Canada–U.S. FTA, in particular those regarding what the agreement is *not* going to achieve. The agreement is not a panacea for all Canada's economic problems. The agreement, by itself and independently of the general economic context and policies, cannot ensure economic development, a more productive and competitive manufacturing sector, the dynamic growth of export-oriented industry or improved welfare. It could potentially make a contribution toward these goals in the presence of reasonable macro- and microeconomic policies and adequate industrial policy measures.* Whether or not this contribution would be major is arguable, and my view is that these benefits would be modest.

A better way to picture the Canada–U.S. FTA is as one element in the overall economic policy of the government. For example, Harris (1985a) supported strongly the concept of a free trade agreement with the United States together with an active industrial policy program to facilitate the adjustment process and encourage the development of new export-led industry. This analysis assumed the use of macroeconomic tools to maintain purchasing power parity in the exchange rates so as not to impair industrial competitiveness during the adjustment period. It also assumed that any agreement would not impair the Canadian government's ability to implement strong trade and industrial policies, as well as independent macroeconomic policies (these assumptions later proved to be false).

According to this economic blueprint, the Canadian government has done it all wrong. There has been no real effort to implement a policy package to promote adjustment in the desired direction. On the contrary, the view was promoted that 'markets by themselves will do the job'. The recessionary policies of high real interest rates and real overvaluation of the dollar have been massively disruptive for the export sector, diminishing the competitiveness of the manufacturing sector to an all-time low, at a critical point in time. There have been very high economic costs of job loss and increased unemployment.

The Canada–U.S. FTA has both costs and benefits; however the economic policies of the government have created a situation where Canada is paying the costs of the agreement but is unable to realize the gains to be had from it. Moreover, this evaluation of costs and benefits has not dealt with the social and political costs of the agreement, nor

---

* 'Reasonable' macroeconomic policies would strive for low real interest rates and purchasing power parity between the Canadian and U.S. dollars

has it dealt with the question of whether the agreement limits the Canadian government's autonomy in areas such as macroeconomic policy, forcing it to promote less than optimal policies. Including these added drawbacks only strengthens the case against the Canada–U.S. FTA.

NOTES AND REFERENCES

1. Harris and Cox's projections of significant free-trade gains for Canada were widely quoted in the specialized literature as well as in the policy debate during the mid and late 1980s, and have now even reached many international economics textbooks. Out of four recent textbooks checked randomly, three cited the Harris–Cox results. See, for example, Ethier (1988, p. 183). A review of quantitative work on the Canada–U.S. FTA can be found in Coughlin (1990).
2. Lazar (1981, Chapter 3) is a good introduction to U.S. contingency protection from a Canadian perspective. Sinclair (1993) presents an up-to-date analysis, with implications for Canada and Mexico. For definition of terms, see the glossary in Grinspun and Cameron (1993).
3. See the qualified evaluation by Donald Macdonald, chair of the Macdonald Commission, in Smith and Stone (1987, pp. 23–5)
4. This paragraph and the next rely on Sinclair (1993). In Article 1907 of the August 1992 NAFTA text the parties agree to consult on a system of rules for countervailing duties and subsidies.
5. Strong support for this argument can be found in Sinclair (1993).
6. I owe this argument to Fred Lazar (personal communication).
7. See Rao and Lemprière, 1992a, p. 3, figure 1, based on data on real output per person-hour, expressed in national currencies.
8. See also John Urquhart, 'Canadian firms, fleeing the high costs at home, relocate South of the border', *Wall Street Journal*, 7 February 1991.

9 The North American Free Trade Agreement: A Canadian Perspective

Edward R. Bruning

9.1 INTRODUCTION

Approximately two years have passed since Canada, Mexico and the United States commenced discussions on an agreement to significantly reduce trade barriers and create a more unified North American market place. Working groups and negotiators for the three parties are involved at this moment in the process of defining the terms and conditions the final agreement will take. One has only to glance at a daily newspaper to catch up with the latest changes that have been whispered to the press.

The more pessimistic observers are apt to predict that the agreement will not be ratified by the U.S. Congress this year. National elections were held which resulted in the election of 110 new Congressmen and 11 Senators. This along with the Clinton administration's vagueness in regards to the extent of its support for NAFTA leads one to the conclusion that ratification of NAFTA will not take place in 1993. Under the 'fast-track' provisions of the trade laws governing international trade agreements by the United States, the President cannot submit the agreement to Congress until he has signed it and he cannot sign it until 90 days after he has given notice of his intent to enter into the agreement. Congress must respond to the proposal after notice has been given by the President. The more optimistic observers point to President Clinton's endorsement of the trade pact and to successful conclusion to its negotiation.

Political uncertainty as to successful ratification of the trade agreement notwithstanding, the purpose of this chapter is to explore the evolving North American Free Trade Agreement (NAFTA) from a Canadian perspective. In approaching this task it is important to establish a firm understanding of the economic situation facing the country in terms of sectoral development and decline, national competitive advantage and Canada's trade and investment relations with both the United States and Mexico. The economic perspective will be presented in Section 9.2. Section 9.3 will address the trade arrangement that Canada established with the United States which culminated in the Canada–U.S. FTA. A concern with many Canadians is that NAFTA is an attempt on the part of the United States to gain concessions it was unable to successfully negotiate for in the Canada–U.S. FTA process. The similarities and differences between the two trade agreements will be assessed. Section 9.4 will outline the specific objectives that Canadian trade negotiators have established in entering into the trilateral process with the United States and Mexico. Section 9.5 will summarize the Canadian position and explore the attitudes and opinions held by various elements of the Canadian polity regarding NAFTA.

Canada is composed of a multiplicity of interest groups who hold quite divergent opinions as to the wisdom of entering into a trilateral trade arrangement. Any attempt to distil the mood of a country on such a high-powered issue as NAFTA in a few short pages is an admitted injustice to all. It is hoped, however, that the message contained in these pages will inspire readers to search out the detailed facts on their own. Much can be learned by viewing the issue from the perspective of those sitting on the opposite side of the table.

9.2 CANADA AT THE CROSSROADS

The fundamental economic goal of any nation is to better the living standards of its people. By this standard Canada has performed quite well since the early 1960s. Table 9.1 provides a comparison of Canada relative to several prominent OECD countries in terms of living standards for 1960 and 1989.

It is evident that the standard of living enjoyed by Canadians ranks among the very best in the world. Since the early 1960s Canadian per capita GDP, private consumption per capita and national disposable income per capita have all improved relative to major OECD countries. These strides in economic achievement have not been at the expense of distributional equity. Canadians pride themselves on their achievements in social/medical care, gender equality, neutrality in international relations and concern for the rights of minority groups. All of these are notable achievements and sources of national pride.

Table 9.1   Ranking among selected OECD countries

Country	GDP per capita		Private consumption per capita		National disposable income per capita	
	1960	*1989*	*1960*	*1989*	*1960*	*1989*
Canada	4	2	4	2	5	4
U.S.	1	1	1	1	1	2
Japan	21	8	21	7	21	7
Germany	11	10	12	14	11	9
U.K.	7	13	5	6	6	11
Italy	18	18	18	11	18	15
Sweden	6	7	8	15	7	8
France	14	11	15	8	13	10

Notes:
(1) Indicators are based on constant prices in US dollars using 1985 PPP exchange rates.
(2) Rankings based on 24 OECD countries.
Source: OECD, *National Accounts*, 1960, 1989.

Past achievements notwithstanding, it is Canada's preparedness for the future that is concerning many of its leaders (D'Cruz and Rugman, 1992; Hart, 1990). Canada exports relatively low-value-added products, particularly metals, forest products, chemicals and water-power electricity. Natural resource exploitation has been the major engine of growth and development. Although Canada is one of the world's largest producers of automobiles and automobile parts, and possesses one of the world's leading telecommunications sectors, its net export of semi-processed and processed, high-value-added exports is rather limited. With the exception of food and beverages, Canada's production of consumption goods and services is low in comparison to world leaders. While Canada's absolute productivity is high, it has the second lowest productivity growth rate over the past three decades among the G-7 nations. Table 9.2 reports growth in GDP per employed person for Canada, the United States, the U.K., Sweden, Germany, Japan, France and Italy.

In particular, Canada has had the lowest labor productivity growth in the manufacturing sector among the G-7 countries (see Table 9.3). It has lagged in machinery and equipment investment (Table 9.4) and R&D expenditure as a percentage of total GDP (Table 9.5). In recent

Table 9.2   GDP per employed persons* and growth rate

Country	Growth Rate		Compound annual growth rate in GDP per employed person 1960–89
	Thousand US dollars 1960	Thousand US dollars 1989	
Canada	18	30	1.9
U.S.	23	35	1.2
U.K.	14	28	2.2
Sweden	13	27	2.2
Germany	11	27	2.9
Japan	5	27	5.3
France	7	21	3.3
Italy	4	17	3.8

*Indicator based on 1985 PPP exchange rates and constant prices.
Source: U.S. Bureau of Labour Statistics.

Table 9.3   Labour productivity growth in manufacturing,* 1979–89

Country	% per year
Japan	5.5
U.K.	4.8
Italy	4.0
France	3.4
U.S.	3.2
Sweden	2.2
Germany	2.0
Canada	1.8

* Growth in real output per hour in manufacturing.
Source: U.S. Bureau of Labour Statistics.

Table 9.4   Machinery and equipment investment as a percentage of GDP, average 1960–88

Country	1960–88	1980–8
Japan	12.1	10.3
Italy	10.0	10.1
France	8.8	8.5
Germany	8.7	8.5
U.K.	8.8	8.4
Sweden	8.5	8.4
U.S.	8.0	8.5
Canada	8.0	8.0

Source: OECD *Economic Outlook*, Historical Statistics.

Table 9.5 Gross expenditure on research and development as a percentage of GDP, 1988

Country	Percentage
Sweden	3.1
Switzerland	2.8
Japan	2.8
Germany	2.8
U.S.	2.8
France	2.3
U.K.	2.3
Italy	1.4
Canada	1.2
Australia	1.2

Source: IMD *World Competitiveness Report*.

years strong reliance on the resource sector has meant declining export earnings since world commodity prices have fallen consistently. The trend in commodity markets suggests a declining attractiveness for some of Canada's traditional export industries (see Figure 9.1).

Figure 9.1 Canada's terms of trade (1970–90) (1986 = 100)

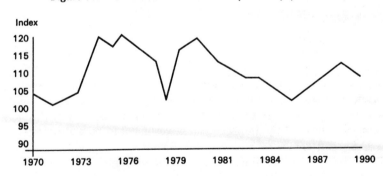

Source: *Statistics Canada*.

Fuelled by chronic federal government deficits, government debt as a percentage of GDP is now the second highest among OECD countries (only Italy's debt is higher). Interest on federal government debt now consumes 4.2 per cent of GDP. In addition, some provinces have also been incurring deficits. Ontario's expected budget deficit has been forecast to exceed $11 billion for 1991 and significant continuing deficits through 1995 to a level of about $70 billion, or approximately 20 per cent of its forecasted GDP. Even in 1990, when federal and provincial debt is added together, Canadian government debt exceeded 70 per cent of GDP. Much of this government debt is increasingly held by foreigners, as the size of the spending requirements coupled with the decreasing Canadian savings rate, have decreased the level of internal funding.

The vital economic signs reported above do not bode well for the country. Declining productivity, falling export earnings and market shares in many important sectors, uncompetitive industrial firms and market structures, and an overspending public sector are signs that change is required. Complicating the matter even more is the fact that many Canadian businesses are inward rather than outward looking, having existed for almost a century in extremely protected markets. Protection, however, holds special relevance within provincial markets. Barriers to interprovincial trade remain high to this day, and not a few business persons refer to a shipment to another province as an export much the same as if the transaction dealt with a shipment across the Canada–U.S. border.

9.3   CANADIAN TRADE POLICY AND THE CANADA–U.S. FTA

The lack of acknowledgement of the extent of global competition and the corresponding need to reassess methods for adapting to the new competitive challenges are the major sources of concern for Canadian businesses. Indeed, the problem can be traced to the National Policy enunciated in 1879 by the Canadian government in part as a response to U.S. indifference to renewing the Reciprocity Agreement of 1854 (Hart, 1990). The National Policy sought to induce economic development and manufacturing in Canada by imposing high tariffs, new infrastructure development and favorable immigration policies. Trade policy therefore evolved as one of import substitution for manufacturing development through high levels of protection while seeking access to foreign markets for Canada's abundant natural resources (Hart, 1990). Even after numerous GATT rounds, Canada has been able to maintain a higher level of protection for industrial products relative to other OECD countries (Waverman, 1991). While a significant watershed was established in the mid to late 1980s with the joint ratification by the United States and Canada of the Free Trade Agreement, Canada entered the 1990s with an industrial structure and domestic manufacturing industry which was small, fragmented and inefficient relative to world competitors.

The often cited justification for the protective trade orientation was that as a small country dependent on trade with larger countries, Canada was in an unequal bargaining position and felt a need to protect itself. While perhaps good negotiating material, the argument ignored two very important realities. First, it denied the basic premises of the open multilateral trading system which Canadian officials were promoting in Geneva. Second, economic policies which shield producers from competition generate substantial welfare losses, particularly when such policies are extended over long periods of time and include a substantial element of potential economic rents garnered through the political process. According to Michael Hart,

> The FTA provided an intellectually respectable basis for undoing this [protectionist] mindset. It is no wonder that some of the strongest initial opposition to the idea of a free-trade agreement with the United States came from officials . . . steeped in Canada's earlier GATT experience (Hart, 1990, p. 45).

The Canada–U.S. FTA has brought about a change in the relationship between Canada and the United States. The Canadians who were most influential in promoting the agreement had three fundamental objectives in mind. First, the agreement was promoted to stimulate domestic economic reform by stifling vestiges of the National Policy and more modern techniques of protectionism. Reduced protectionism would require Canadian businesses to inculcate a more global view when establishing competitive strategies and, consequently, would force them to modernize their equipment and processing techniques to meet present-day competitive demands. Second, the Canada–U.S. FTA was Canada's safeguard for a growing mood of U.S. protectionism. Establishing a bilateral deal with its cousin to the south provided Canada an open and secure entry into the larger U.S. economy. Third, the agreement provided a more appropriate mechanism for managing the Canada–U.S. economic relationship. The original GATT provisions have proven inadequate in rectifying several of the major sources of contention between the two countries. More enforceable rules and a suitable institutional infrastructure would put the relationship on a more predictable basis (Hart, 1990).

9.4   CANADIAN–MEXICAN–U.S. TRADE AND INVESTMENT PATTERNS

9.4.1 Trade

Canada and the United States have had a long history of reciprocal trade. In fact, in total volume the trade between the two countries is larger than any other nation-pair in the world. Tables 9.6, 9.7 and 9.8 reflect the geographical distribution of merchandise trade for Canada, the United States and Mexico, respectively.

Table 9.6   Geographic distribution of Canadian merchandise trade (percentages)

| | 1988 | | 1989 | |
Trading Partner	Exports	Imports	Exports	Imports
U.S.	73.0	65.7	73.3	65.2
Mexico	0.4	1.0	0.5	1.3
European Community	8.0	12.2	8.5	11.1
Japan	6.5	7.0	6.5	7.1
Other Latin America	1.4	2.0	1.1	2.1
Rest of the world	10.8	12.0	10.1	13.3

Source: *Statistics Canada.*

Table 9.7   Geographic distribution of merchandise trade (percentages)

| | 1985 | | 1987 | |
Trading Partner	Exports	Imports	Exports	Imports
Canada	22.0	20.0	23.7	17.5
Mexico	6.4	5.5	5.8	5.0
European Community	21.5	18.8	24.0	20.0
Japan	10.6	19.9	11.2	20.8
Other Latin American	6.7	7.0	6.7	5.9
Rest of the World	32.7	28.8	28.7	30.7

Source: U.S. Department of Commerce, *Statistical Abstract of the United States*, 1987 and 1989.

Table 9.8   Geographic distribution of Mexico's merchandise trade
(percentages)

Trading Partner	1985		1987	
	Exports	Imports	Exports	Imports
United States	61.2	67.9	64.7	6.6
European Community	10.3	13.2	14.6	16.2
Japan	7.9	5.6	6.5	6.5
Latin America	5.8	4.8	7.5	2.9
Canada	1.7	1.6	1.5	2.7
Rest of the World	14.2	7.0	5.1	7.0

Source: Adapted from Sidney Weintraub, *Marriage of Convenience*, tables 4.1 and 4.2.

Evident in the data is the fact that both Canada and Mexico secure most of their export earnings from the United States. Both Canada and Mexico rely on U.S. markets for approximately 65–75 per cent of exports, while total imports from the United States average about 65 per cent.

Canada and Mexico have remarkably similar trade patterns as the United States. At the same time, neither country is significant in the other's trading pattern.

Table 9.9 presents a more detailed breakdown of trade flows between the countries.

Trade is classified into four categories: resource-intensive products (animals, beverages, crude materials and so forth); primary manufacturing (manufactured goods by material – iron and steel, aluminum smelting and so forth); the machinery and transport sectors; and all other secondary manufacturing (Waverman, 1991).

In 1988–9 approximately 47 per cent of Canadian exports and 33 per cent of Mexican exports to the United States consisted of resource-intensive or primary manufactured products. Nearly 60 per cent of Canadian exports to Mexico were resource or primary manufactured products, while nearly 70 per cent of Mexican exports to Canada consisted of secondary manufactured products; 72 per cent of U.S. exports to Canada and Mexico were from secondary manufacturing. The pattern that emerges is one of Canadian exports of resources and primary and secondary manufactured materials being exchanged for other primary materials, but mainly for secondary manufactured

Table 9.9   Canadian merchandise trade with Mexico (Can.$ 000s)

Product	1988		1989	
	Exports	Imports	Exports	Imports
Food	157845	112206	150336	112354
Beverages and tobacco	307	12596	223	15363
Crude materials, inedible	48547	69172	45220	23295
Mineral fuels	2886	59753	38	49406
Oils and fats	1827	—	1741	—
Chemicals	18872	13505	7668	13845
Manufactured goods	117261	191064	161829	332104
Machinery and transport equipment	129798	828706	213830	1100863
Miscellaneous	3954	29965	4769	42381
Other articles	7662	10716	17397	8,718
Total	489002	1327726	603098	1698368

Source: *Statistics Canada.*

goods. Mexican exports of secondary manufactured goods represent goods with a high component of low-skilled assembly labor.

Waverman (1991) has characterized trade between Canada and the United States as intra-industry, with similar technology and uses of factors of production in the two highly developed economies. Trade between Mexico and Canada, however, is inter-industry in nature, reflecting sharp differences in technology and productive factors.

Many Canadians believe a free trade agreement between Mexico, the United States, and Canada would expand the North American trade pattern to predominantly one of high-labor-content secondary manufactured goods from Mexico, with resource products being exported from Canada to the United States and Mexico. Many believe Canada could never compete with a Mexican–U.S.-dominated secondary manufacturing sector and that NAFTA would wed them to a resource-dependent export strategy (Globerman, 1991). On the contrary, those less concerned with past trade relations argue that history would not have to repeat itself if Canadian manufacturers were to specialize in activities that employ high-skilled labor inputs matched with sophisticated technologies. The real issue Canadians must address is how to increase labour and capital productivity in order to take advantage of the bountiful resource endowment possessed by its well trained labor force (Waverman, 1991).

9.4.2 Investment

At the end of 1989, the estimated accumulated FDI in Mexico stood at approximately U.S.$26.5 billion. By country of origin, 95 per cent of the stock of foreign investment came from ten countries. The United States is the largest direct investor in Mexico by a decisive margin. Although the UK and Germany saw their relative shares increase marginally during the 1980s, the respective positions of other foreign investors in Mexico have not changed substantially. Canada's share of FDI in Mexico hovers around 1.4 per cent and is the eighth largest source of foreign capital for Mexico.

Within manufacturing, the bulk of U.S. direct investment in Mexico is concentrated in three major industries, namely chemicals, transportation equipment and 'other' manufacturing. These industries together account for almost one-third of all direct investment in Mexico. U.S. direct investment in Mexico's transport industry is of particular concern to Canadian automobile manufacturers. The primary motivation for such investment centres on the significant cost advantages in Mexico's maquiladora region. The 'big three' U.S. automobile producers were among the top five of the five hundred largest enterprises in Mexico in 1987, with 100 per cent U.S. ownership in the Mexican affiliates. The tax and duty advantages attached to maquiladora operations could significantly affect Canadian automobile interests in a negative way.

While U.S. FDI dominates foreign capital flowing into Mexico, a similar dominance is experience in Canada as well. At the end of 1989, U.S. FDI in Canada amounted to roughly U.S.$67 billion, the largest concentration of American FDI in any host country. Although Canada is the largest recipient of U.S. direct investment, the total share of U.S. direct investment abroad flowing to Canada fell from approximately 30 per cent in the 1950s to approximately 19 per cent in 1992. The establishment of the Foreign Investment Review Agency (FIRA) and the National Energy Program discouraged U.S. foreign investment in Canada. With the liberalization of the foreign investment rules brought about by the passage of the Investment Canada Act in 1985, which replaced the more restrictive FIRA, U.S. direct investment in Canada has accelerated (Hart, 1990).

Although the nationalistically inspired FIRA has been superseded by a more internationally oriented investment regime, U.S. FDI is still a major irritant to many Canadians since U.S. capital dominates so large a portion of Canadian industry. By the end of 1989 the largest proportion of U.S. direct investment in Canada was concentrated in manufacturing (48 per cent), the financial sector (19 per cent) and petroleum (16 per cent), with the remaining investments spread broadly across numerous 'other' industries and sectors. The 'other' sectors are composed mainly of natural-resource-based industries, construction and the retail trade. Since 1984 U.S. direct investment in the petroleum industry has declined from 6 per cent to 4.4 per cent. A major impetus for the reduction is directly associated with implementation of Canada's energy policies, which called for an increase in Canadian participation in the industry. Within Canada's manufacturing sector, U.S. direct investment has traditionally flowed to the transportation sector, primarily motor vehicle and equipment manufacturing. However, in the late 1980s, the 'other' manufacturing sectors received almost 60 per cent of the increase in U.S. FDI.

An interesting trend that appears to be emerging is a decline in the importance of U.S. FDI in automobile manufacturing in Canada relative to other countries, particularly Mexico. Recognition of the shift in economic activity in the auto and other core manufacturing industries has generated intense concern about the future of Canadian jobs. Automobile production is a major source of employment in Canada, and a trade pact which threatens jobs in a major employment sector is sure to meet a serious challenge from labor organizations and the political left.

9.5 CANADIAN NATIONAL UNITY DEBATE

Intense discussions are under way among groups in Canada to determine how best to deal with the 'new' economic realities of a North American integrative initiative – serious discourse is occurring in town halls, municipal centres, and provincial and national offices throughout the country as to whether Canada will indeed remain a single nation or split into two parts, one composed of Quebec and the other the rest of Canada. The debate has raged for as long as the nation has existed; however, with its growth in affluence and political strength, French-speaking Canada has called for a reassessment of its interest in remaining a party to the original confederation. The factors that separate the two groups are not so much economic as they are political in nature. The leadership of French speaking Canada is committed to preserving the language and culture of its constituents, apparently at all costs, although the economic realities of an actual breakup are encouraging many would-be sympathizers to reconsider their initial positions. The French-speaking leadership called for a referendum in 1992 to determine whether Quebec should remain part of Canada. English-speaking Canada, sections of which were largely indifferent to the outcome of the issue, responded minimally to the threats of secession by Quebec. The general tenor of their responses was 'let them go their way and we ours'.

A lull has occurred in the national unity debate in Canada. In October 1992 Canadians overwhelmingly defeated a proposed constitutional change that would grant special status to Quebec, among other important changes that affected native peoples, the structure of the national senate, and the rights and privileges of distinct minorities. The defeat of the proposed constitutional changes, however, is not necessarily a reflection of the general sentiment of the rest of Canada's view of Quebec. More than half of the provinces, including Quebec, refused to support the proposed constitutional changes. Quebec, and the rest of Canada, are in the process of determining their positions on the specific issues brought out in the recently defeated referendum, and debate is expected to continue in the not too distant future.

With respect to free trade and NAFTA, very little separates French- and English-speaking Canadians. Pro-free-traders are equally represented on both sides. Quebec's outward trade orientation has always been strong, even though the willingness to assimilate culture traits from others has been fought with vehemence. Similarly, a substantial, although not necessarily a majority, of English-speaking Canadians support the free trade and North American integration initiatives. Most importantly, however, the Tory government which presently wields power in Ottawa, even though its popularity is at the lowest level of any Canadian government in this century, is a strong advocate of free trade and economic integration. Its efforts to effect changes in attitudes and perceptions towards support of NAFTA have been seriously undermined by attention being focused on national unity issues. The seriousness of the debate to the future of Canada could not be greater. All societies at one point or another must face what it is they are and what they want to become. It is unfortunate, however, that Canada's search for a national identity must occur at the same time as tremendous economic transformations are taking place that require a focusing of priorities and energies in order for the country to negotiate and plan effectively for the future.

9.6 NAFTA AND CANADIAN ECONOMIC INTERESTS

The Canadian approach to the NAFTA negotiations builds on the Canada–U.S. FTA and is closely coordinated with Canada's efforts in the Uruguay Round of Multilateral Trade Negotiations (MTN). For a number of areas such as agriculture, subsidy/countervailing duties and intellectual property issues, Canada believes that the best chance of progress lies in a successful conclusion to the Uruguay Round. Canada's broad objectives in the NAFTA negotiations are:

1. Barrier-free access to Mexico for Canadian goods and services, while developing tariff phase-out provisions and safeguard mechanisms which reflect Canadian import sensitivities.
2. Improved security of access to the U.S. market (clearer rules of origin; improved customs cooperation).
3. Improved conditions under which Canadian businesses can make strategic alliances within North America to better compete with the Pacific Rim and the European Community.
4. Ensuring that Canada remains an attractive site for foreign and domestic investment.
5. The establishment of a fair and expeditious dispute settlement mechanism.

It is important for Canadian negotiators that the U.S. team does not view the comprehensive NAFTA as an extension or expansion of the Canada–U.S. FTA, which is specifically designed to fit the special relationships and interests between Canada and the United States. The principles behind the Canada–U.S. FTA, though, are expected to help guide the NAFTA negotiations. The two agreements must be complementary in many important respects – including rules of origin and dispute settlement provisions. Every effort must be made to avoid the so-called hub-and-spoke approach, whereby bilateral arrangements between the United States and Canada, and between the United States and Mexico, would result in an overall net advantage for the United States. Recognizing the merits and shortcomings of a free trade agreement between a low-wage, developing country and two industrialized, high-wage countries, Canada believes that the agreement between the United States and Mexico is inevitable and wishes to participate in order to represent its own interests. By helping forge the conditions by which trilateral free trade can be introduced, Canadian sectoral impacts can be anticipated and addressed in more effective ways. The phasing in of tariff reductions, exemptions for certain sectors, rules of origin, protection of industrial support programs and other components of NAFTA can be designed to ease sectoral impacts. Work on tariff and non-tariff barriers is of vital interest to Canada. It is important that tariff phase-outs are structured in terms of immediate, intermediate and longer-term stages. While recognizing the developmental status of the Mexican economy, and while cognizant of the longer term gains to all parties that would be brought about by Mexican economic prosperity, Canada is concerned with assuring an equitable phasing-in of more liberalized trade.

Rules of origin are outlined in the Canada–U.S. FTA, and Canada is interested in the effects of including a third party in the trade relationship. Canadian negotiators are interested in assuring that rules of origin are such that benefits of free trade accrue to NAFTA participants on an equitable basis. At the same time, it is paramount that the certification process does not itself become a burden to liberal trade between the three countries.

Canada has pressed both the United States and Mexico to liberalize their government procurement regimes, using the GATT Procurement Code plus the Canada–U.S. FTA commitments as the starting point for further improvements. Included as an item to be negotiated, and one which Canada would like to see implemented, is the inclusion of services procurement in the agreement. Canada is interested in liberalizing the purchasing practices of Mexico's parastatal (government corporations) which account for in access of U.S.$8.1 billion of procurement business annually.

With respect to agriculture, experience under the Canada–U.S. FTA indicates that while bilateral, or regional, progress can be achieved, several of the core problems in international agricultural trade can only be fully addressed at the multilateral level. Thus Canada continues to emphasize the importance of achieving a balanced result in the Uruguay Round. In the NAFTA context, there has been progress made with respect to rules of origin and tariff reductions. Discipline on the use of export subsidies is also a key element in the Canadian negotiating agenda. Also of concern are measures dealing with sanitary and environmental issues, non-tariff measures and a special agricultural safeguard provision.

Canada shares Mexico's desire to see greater discipline on trade-remedy procedures, particularly subsidy/countervailing and antidumping determinations, but believes that priority at this point should be given to the MTN. Canada's objectives in the area of safeguards are to develop an emergency safeguard mechanism which would improve Mexican procedural and institutional standards of transparency and to craft a transitional provision capable of addressing effectively any Canadian import sensitivities.

With respect to intellectual property rights, Canadian negotiators are fully aware that any progress on this issue will be heavily influenced by MTN successes. Canada contends that both agreements should contain strong enforceable provisions requiring adequate standards of protection for intellectual property rights.

A number of additional issues are on the agenda to be discussed/ negotiated by the respective national teams. Investment issues and the liberalization of service sectors and, particularly, financial services and insurance are high on the list of topics to be addressed. One of the more sensitive issues that is only indirectly related to trade and investment is the impact the proposed NAFTA will have on the environment. Numerous interest groups throughout Canada have stated in clear and unequivocal terms the need for the negotiating team to uphold current environmental standards in the developed regions of North America. Thus, negotiators are implored to negotiate for consideration of trade-related environmental impacts and the setting of standards, technical regulations and procedures for products affecting health and human welfare. Canadian negotiators are stating the case that each country must maintain the ability to set environmental standards and to enact measures necessary to ensure compliance with internationally agreed environmental accords.

9.7 POLITICAL SENTIMENTS

The NDP, Canada's equivalent to European social democratic parties, has gone on record as strongly against ratification of NAFTA. In its mind, a trilateral trade pact is another American multinational trick to obtain a greater portion of Canadian wealth at low cost. The Liberal Party, while less vociferous than the NDP in its criticism of Tory internationalism, nonetheless has not spoken out in favor of the trade agreement. Rather, the leadership has gone on record as being in favour of rescinding the Canada–U.S. FTA if restored to national power. Both the NDP and the Liberal parties have courted labor and working-class voters throughout the century. Rank and file voters, with an eye to their pocketbooks as opposed to some abstract ideological construct, when pressed to express their political views have publicly spoken out against NAFTA.

Various citizen action groups and other splinter elements of Canadian society have adopted a stance that is quite unsympathetic to NAFTA. It is rather more difficult to ascertain whether or not the idea of an agreement is supported by the citizenry as a whole. Latent and not so latent anti-Americanism, fear of third-world-country disregard for environmental quality, a concern that low-wage workers would be exploited, and substantial discontent with present-day Canadian political and economic institutions has resulted in a rather messy view of NAFTA proposals. Most people are unaware of the ramifications and objectives of the NAFTA; however, most are likely to state whether they are for or against the proposed agreement. On the other hand, business leaders by and large are cognizant of the stakes involved and are strong endorsers of the pact. The Tory leadership has steadfastly promoted free-trade since taking national office in 1984. Together with the prosperity and competitiveness initiative, the government's free trade program, as represented in the Canada–U.S. FTA and proposed NAFTA, offer the country a means to redirect and initiate energies to enhance productivity performance in numerous sectors of the economy. Canada has both tremendous potential and the natural and creative resources to remain a major force in North America. Decisions rendered over the next several years in the political and economic realms will determine the future for the country's role in North America.

10 Mexico in a North American Free Trade Area

Joseph A. McKinney

10.1 INTRODUCTION

On 5 February 1991 the leaders of Mexico, the United States and Canada simultaneously announced that their governments would enter negotiations to form a trilateral free trade area. The proposed North American Free Trade Agreement (NAFTA) will encompass a population of 360 million people with a total product of more than $6 trillion. It aims toward unrestricted trade in goods and services among the three countries, but not a common external tariff or unrestricted labor mobility as in the European Community.

A number of events had set the stage for these trilateral trade negotiations. The United States approved a free trade agreement with Israel in 1984 and stated that requests from other nations for similar agreements would be considered. In 1985 Canada requested such an agreement with the United States. The Canada–U.S. FTA was negotiated in 1987 and, after a bitterly contested national election on the issue, went into effect on 1 January 1989.

Canada had entered into free trade negotiations with the United States with some apprehension, fearing that closer ties with the United States might weaken Canada's sense of cultural identity or compromise its national sovereignty. But Canada had observed rising protectionism in the United States during the 1980s. The United States, experiencing large and persistent balance of trade deficits but being constrained from raising tariffs or imposing quotas by the disciplines of the General Agreement on Tariffs and Trade (GATT), had resorted to voluntary restraint agreements and 'administered protection' in the form of more frequent use of antidumping and countervailing duty cases.

Although Japan was the primary target of United States trade policy actions, Canada and other countries were increasingly affected by them. Canada viewed the free trade agreement as a way to assure access to the U.S. market, a market which accounts for about 70 percent of

Canadian exports. Another goal of free trade proponents in Canada was to force a restructuring of Canadian industry to make it more competitive internationally (Hart, 1990, p. 23). The Canada–U.S. FTA provided for the gradual elimination of all import tariffs, the removal of many nontariff barriers to trade, liberalization of investment regulations and the establishment of a bilateral dispute settlement mechanism.

10.2 ECONOMIC AND TRADE REFORMS IN MEXICO

While Canada was considering a free trade agreement with the United States, dramatic changes were taking place in Mexico. These changes would eventually lead to a restructuring of the Mexican economy and a complete reversal of its trade policies.

During the post-Second World War period, Mexico's trade policy had been one of restricting imports in various ways to encourage import substituting industrialization. High import tariffs, official import reference prices, domestic content requirements, subsidies and import licenses were among the instruments used to discourage imports. Foreign investors often had to agree to export a certain percentage of their output from Mexico.

For many years Mexico had resisted becoming a contracting party to the GATT, fearing that acceptance of GATT's principles might compromise Mexico's import-substituting industrialization strategy. GATT membership was considered but rejected in 1980. Those benefiting from trade restrictions in Mexico joined forces with nationalists who feared that more liberal trade policies would threaten Mexican autonomy to defeat the proposed membership.[1]

In 1980, when Mexico rejected GATT membership, the country's export earnings were inflated by the high price of petroleum, with petroleum sales accounting for almost 80 per cent of total export earnings. Despite these inflated export earnings Mexico had large balance of payments deficits which it financed by external borrowing. Mexico's current account deficit grew tenfold between 1970 and 1981, increasing from 3 per cent to 5 per cent of GDP (Weintraub, 1988, p. 9). Optimistic projections of future oil revenues made it possible to finance these deficits through external borrowing.

During this same period, Mexico's fiscal deficit increased from less than 3 per cent of GDP to over 17 per cent. In some years almost two-thirds of total government expenditure went to subsidies or other support of domestic industries (USITC, 1990a, p. 1.2). The share of public sector expenditure in GDP increased in Mexico from 14.6 per cent in 1975 to 25.6 per cent in 1983 (Sobarzo, 1992, p. 606).

The decline of the crude oil price in 1982 presented Mexico with a serious economic crisis. Current account deficits could no longer be financed by external borrowing. Since the accumulated foreign debt had to be financed out of net export earnings, imports were slashed from just under $24 billion in 1981 to $8.5 billion in 1983. Mexico's current account balance changed dramatically, from a $12.5 billion deficit to a $4.5 billion surplus, during the same period (Weintraub, 1988. p. 10).

President Miguel de la Madrid's administration began a gradual trade liberalization in 1983 to encourage restructuring of Mexican industries. An executive decree in July 1985 significantly increased the tempo of trade liberalization and marked the beginning of a profound shift of trade policy by Mexico. Whereas in 1982 all products imported into Mexico had required an import license, this decree removed the import licensing requirement from almost two-thirds of all products (USITC, 1990a, pp. 2.2–2.3)

In 1986 Mexico reversed its earlier position and joined GATT. As one of the conditions of accession to GATT, Mexico agreed to bind its entire tariff schedule to a maximum ad valorem rate of 50 per cent, and in fact reduced tariffs to a maximum rate of 20 per cent and a weighted average rate of only 11 per cent. Import licensing requirements were further liberalized, official reference prices for imports were virtually eliminated, and Mexico agreed to sign five of the nontariff barrier codes of the Tokyo Round (USITC, 1990a, pp. 2.2–2.3)

The Economic Solidarity Pact announced by the Mexican government in 1987 contained further trade liberalization measures. This plan involved a wide-ranging set of reforms aimed at lowering the inflation rate (which by that time was almost 200 per cent) and increasing productivity. Austerity measures between 1982 and 1987 had caused per capita incomes to decline by 15 per cent, with an even greater decline in the real incomes of wage earners (USITC, 1990b, p. D.2)

The economic reforms of the Economic Solidarity Pact moved Mexico further in the direction of market-determined resource allocation and outward-looking trade and industrialization policies. Policy actions by the government continued reducing import licensing requirements and import tariffs (to a maximum 20 per cent rate), eased foreign investment regulations and laid the plans for privatization of state-owned industries. The reform process continued into 1992 with the further removal of trade restrictions, deregulation of some sectors of the economy and the sale of over $14 billion of state-owned assets and the transfer of more than 250,000 workers from the public sector to the private sector (Rogozinski, 1992, p. A11).

10.3 UNITED STATES–MEXICAN TRADE RELATIONS

An important change in United States–Mexican trade relations took place in November 1987 with the signing of the 'Framework of Principles and Procedures for Consultation Regarding Trade and Investment Relations'. The two countries previously had no formal mechanism to govern their trade relations. In this agreement they set forth certain governing principles for both trade and investment relations, with the hope that this structure would help to alleviate the tension and mistrust which had often characterized the bilateral relationship (Erb, 1989).

Rather than extending trade concessions this agreement established a consultative mechanism in which the two countries agreed to discuss within 30 days any trade or investment matter that either country wished to discuss. They also agreed to begin negotiations within 90 days on a number of issues of immediate concern to the two nations. These negotiations led to agreements which reduced trade barriers on steel, alcoholic beverages and textiles.

Trade officials also held consultations on such matters as intellectual property protection, investment restrictions, agricultural trade, motor transport, the Generalized System of Preferences, and unfair trade cases pending against Mexico (USITC, 1990a, pp. 2.3–2.4) These consultations had a salutary effect on United States–Mexican political relations by reducing trade frictions between the two countries.

Another step forward in the evolving trade relationship between the United States and Mexico was taken when, during a visit of Mexican President Carlos Salinas to the United States in October 1989, an 'Understanding Regarding Trade and Investment Facilitation Talks' was signed. Among other things this agreement provided that information to be used in bilateral negotiations be collected and analyzed by binational teams of experts so that technical issues and differences of interpretation could be settled prior to the negotiations (USITC, 1990a, p. 2.6).

Trade officials from the two nations identified in November 1989 certain topics for possible negotiation, and a decision on whether to initiate these negotiations was to have been made in March 1990. Instead, at a meeting between U.S. and Mexican trade officials in February 1990, the discussion focused on U.S. statutory requirements for entering into negotiations for a comprehensive free trade agreement. Mexico subsequently requested such an agreement, Canada decided that it should be a part of the negotiations to protect its own interests, and negotiations among the three countries began in June 1991. An agreement was reached in August 1992, but must be ratified by the governments of all three countries before going into effect.

10.4 IMPORTANCE OF NAFTA TO MEXICO

Participation in NAFTA is extremely important for Mexico. Since Mexico's trade with the United States accounts for roughly two-thirds of its foreign trade, anything that significantly affects this trading relationship has profound implications for Mexico. Through trade liberalization and economic reforms Mexico is counting heavily on the presumption that more market-oriented and outward-looking economic policies will stimulate economic progress. But the success of these policies depends heavily upon Mexico's having secure access to the U.S. market.

Mexico already has enjoyed relatively free access to the U.S. market, but the permanence of that access has been uncertain enough to inhibit some of the investment which Mexico needs so much for modernization of its economy. In recent years the United States has been responsible for a large percentage of the voluntary export restraints that have been imposed, and for the preponderance of the world's antidumping and countervailing duty cases (USITC, 1990b, p. D.2). A recent GATT report noted that the majority of disputes coming before it involve the three nations of North America (Morton, 1991). Codification of a commitment to free trade within North America and the establishment of an effective dispute settlement mechanism is viewed as a way to increase investor confidence in Mexico.

Several computable general equilibrium models have been constructed to estimate the effects on Mexico of free trade with the United States.[2] The estimates of these models vary widely depending on their assumptions. For example, the results are sensitive to whether or not the models allow for the removal of nontariff trade barriers as well as tariffs, whether or not they consider the effects of capital flows and labor migration, and whether or not they allow for imperfect competition and economies of scale. In general, however, the models estimate that the effects on Mexico from participation in a NAFTA will be strongly positive.

Mexican economic reforms and the prospect of a free trade agreement with the United States have already affected Mexican capital inflows. Gross capital flows into Mexico amounted to more than $22 billion during 1991 (Fidler, 1992, p.4). The value of shares traded on the Mexican stock market more than doubled in value during 1991, and Mexican firms increasingly used international capital markets to raise funds.

The goal of the Mexican government is to attain single digit inflation during 1992. That goal probably will not be attained, and the contractionary measures imposed upon the economy by the government in trying to attain the lower inflation rate have slowed the rate of economic growth. Mexico's current account deficit has widened rapidly, but this is to be expected when there are large capital inflows and a demand for goods which has built up over several years of austerity. More than three-fourths of Mexican imports currently are intermediate and capital goods. Since these goods will increase future productive capacity, the borrowing from abroad is not a matter of concern for the future.

Mexico's current account deficit is affected by the fact that exchange rate policy in Mexico is causing an increasingly overvalued peso. Inflation in Mexico is approximately 12 per cent, and in the United States it is less than 4 per cent. Yet the peso is being allowed to depreciate at only 2.5 per cent per year, and this implies a real appreciation of the peso. Admittedly, productivity is growing a little more rapidly in Mexico than in the United States, but not enough to keep the peso from appreciating in real terms relative to the dollar.

Maintaining stability in the real value of the exchange rate will be important to the smooth functioning of NAFTA. A persistently overvalued peso could damage the nascent export industries that are being established in Mexico to take advantage of the open markets of the United States and Canada. This in turn could cause disenchantment with the free trade agreement and a retrenchment from its provisions. Conversely, undervaluation of the peso could be problematic. Labor and environmental groups in the United States and Canada have worried about adverse social harmonization pressures coming from Mexico due to the lower labor standards or environmental standards there. Economists contend that such differences generally do not lead to harmonization pressures so long as the exchange rate is free to adjust. Therefore, until such time as social policies among the countries converge, avoiding undervaluation of the peso is important to prevent undue social harmonization pressures in the United States and Canada.

The permanence of economic reforms in Mexico may be importantly affected by whether or not a NAFTA is approved. President Carlos Salinas de Gortari has boldly taken great risks to push forward the economic reform process. The Mexican people have been remarkably patient in enduring economic hardship while the economy was being restructured. Without rapid economic progress there is some danger of retrenchment from economic reforms in Mexico, and economic progress is much more likely if the Mexican economy is economically integrated with the other North American economies.

10.5 IMPORTANCE OF MEXICO TO NAFTA

In comparison with the United States and Canada, Mexico is relatively small in terms of both land area and GDP. The Mexican economy is less than half as large as that of Canada, and only about 4 per cent as large as that of the United States. However Mexico has a population of 86 million people which is growing rapidly and is expected to reach 100 million by the turn of the century. As Mexico's economic reforms lead to an increase in the economic growth rate of the country, market potential will also increase rapidly. Export prospects for U.S. and Canadian firms are particularly good for such products as telecommunications equipment, industrial machinery, transportation equipment, chemicals and allied products, computer systems and software, medical equipment, oil and gas field equipment, certain agricultural products such as wheat and corn, and processed food products.

Because of the relatively small size of the Mexican economy, the short-term economic impact of incorporating Mexico into NAFTA will be relatively modest. Trade with Mexico accounts for only about 7 per cent of total U.S. trade. Mexico's trade with Canada is minuscule, amounting to just slightly over $2 billion per year. General equilibrium models have estimated positive, but almost insignificant, effects upon both the U.S. and Canadian economies from a free trade agreement that incorporates Mexico.

The effects of the free trade agreement are likely to be more significant than the econometric models predict, however, for several reasons. The migration of capital from Canada and the United States to Mexico will cause some contraction of labor-intensive industries in both countries, with attendant adjustment problems unless trade liberalization is phased in gradually, and unless effective trade adjustment assistance programs are in place. The stimulus to exports from economic progress in Mexico should assure that there will be no *net* loss of employment in either the United States or Canada. In fact, empirical work by Gary Hufbauer and Jeffrey Schott, among others, has indicated that the United States can expect net job creation from the free trade area (1992). Nevertheless, the *composition* of employment will undoubtedly change to some degree in both countries as a result of the free trade agreement. Unskilled and semi-skilled labor may lose relative to skilled labor and capital owners unless measures are implemented to augment the skills of low-skilled workers. Such measures would be prudent for both the United States and Canada even in the absence of a North American free trade agreement, for the ongoing international integration of economies will require labor adjustment in any event.

There will be dynamic gains from free trade, and these are also difficult to capture in econometric models. Economies of scale made possible by the trinational market will increase productive efficiency in all three countries. There will be a stimulative effect from increased competition in the combined markets, and a more efficient market structure as monopolistic or oligopolistic market structures are undermined. And the greater certainty of market access in a free trade area will make possible a more efficient investment pattern in each of the countries.

The immediate role that Mexico will play in NAFTA, in addition to being an important market for some U.S. and Canadian goods, will be to serve as a low-cost labor source for U.S. and Canadian firms. Japan has invested extensively in Asia in order to shift labor-intensive production processes to those countries with lower labor costs. The Western European countries now have access, by virtue of the economic and political reforms in Eastern Europe, to a large and productive pool of relatively low-cost labor in those countries. In order for U.S. and Canadian firms to remain competitive, they will need to shift the labor-intensive parts of the production process to labor-abundant areas.

With its abundant labor supply Mexico is an ideal location for the labor-intensive production processes of U.S. and Canadian firms. The Mexican propensity to consume North American products will have much more favorable repercussions than if the offshore investments were located in Asia, where the propensity to consume U.S. and Canadian products is much lower.

Environmental groups have expressed concern about incorporating Mexico into NAFTA. Environmental regulations are indeed less stringently enforced in Mexico than they are in either Canada or the United States. If a free trade agreement were to increase industrial production in Mexico without any improvement in the enforcement of environmental standards, the environment could suffer.

Areas of the United States bordering on Mexico have legitimate concerns about both air and water pollution spilling over from Mexican border cities into U.S. border cities. These issues need to be addressed, and by drawing attention to them the free trade agreement makes it more likely that solutions will be found. Environmental degradation in Mexico does not result from a lack of will on the part of Mexico to address environmental issues, but rather from a shortage of resources to deal with environmental problems. Since NAFTA can be expected to benefit the Mexican economy, the long-term implications for the environment should certainly be positive.

As incomes rise in Mexico, the problem of the illegal immigration of Mexican workers into both the United States and Canada will eventually diminish.* Emigration from Mexico is inversely proportional to the level of economic prosperity there, rising when economic conditions in Mexico deteriorate and falling as economic conditions improve. Social problems arising from illegal immigration will be reduced proportionately as immigration reduces. Wisely, the free-trade agreement proposes a fifteen year phase-in for the trade liberalization of sensitive sectors of Mexican agriculture. This will provide time for industrial employment expansion to provide employment alternatives for labor displaced from agriculture, and thus should prevent an otherwise large migration of labor from rural Mexico to urban areas of both Mexico and the United States.

---

* Writing about the tendency for convergence of productivity and income levels of the United States and Mexico, Clark Reynolds states that 'Convergence can take place through migration, trade, investment, or technology transfer. But if the latter three possibilities are constrained, then migration becomes the main channel through which pressures for convergence operate' (Reynolds, p. 242). The impact of free trade on labor migration in the short term is uncertain, since increased economic activity in the northern region of Mexico may attract more workers there, some of whom them will continue to migrate further north across the border. But the long term effects of free trade on the Mexican economy should, by improving economic conditions there, reduce the incentives for emigration.

Finally, there should be definite political gains from incorporating Mexico into a free trade agreement. Closer economic ties will lead to stronger political ties. A stable and prosperous Mexico, for which the free trade agreement will improve the prospects, is certainly in the self-interest of both the United States and Canada. In addition, a free trade agreement with Mexico is a likely precursor to the hemispheric free trade area which President George Bush proposed in his Enterprise for the Americas Initiative. Already, Mexico's approach to the United States has been a catalyst for several other free trade proposals in Latin America. Brazil, Argentina, Uruguay and Paraguay have announced plans for a Southern Cone common market by the end of 1994. Chile, Venezuela, Colombia and Mexico have announced that they will initiate free trade talks. Mexico and the Central American countries hope to have a free trade area by the end of 1996. The moribund Central American Common Market seemingly is coming back to life. Chile has approached the United States about the possibility of a free trade agreement. All of these developments bode well for closer Western Hemispheric economic and political relations.

10.6 CONCLUSION

Incoroporating Mexico into NAFTA has the potential for benefiting not only Mexico, but the United States and Canada as well. Because of the small size of Mexico's economy relative to those of the United States and Canada, the economic impact of a free trade agreement will be much greater for Mexico than for either of the other two countries. Beyond the immediate impact of reducing trade restrictions, a free trade agreement is important to Mexico because it will help to ensure the permanence of the economic reforms and trade liberalization measures already taken there. In addition, such an agreement will stimulate capital inflows into Mexico, both from foreign investors and from repatriated Mexican flight capital.

While the economic benefits of a free trade agreement will be less significant for the United States and Canada, these benefits will not be inconsequential. Increased production sharing, with the labor-intensive parts of the production process being carried out in Mexico where labor costs are much lower, will increase the international competitiveness of U.S. and Canadian industries. In addition, as Mexico prospers and modernizes both its industry and its social infrastructure, demand will increase there for the types of products in which the

United States and Canada have comparative advantage. Finally, Mexico's incorporation into NAFTA could serve as a catalyst for closer economic integration and political cooperation throughout the Western hemisphere.

NOTES AND REFERENCES

1. For an interesting and detailed discussion of the reasons for Mexico's decision not to enter GATT in 1980, see Weintraub, 1984, pp. 84–91.
2. A number of these models are described, and their results presented, in USITC, 1992.

11 Facing up to Mexico*

Peter Morici

11.1 INTRODUCTION

In September 1990, President Carlos Salinas de Gortari informed President Bush that Mexico would like to negotiate a free trade agreement. The following June, after much heated debate, Congress granted the president fast-track negotiating authority.

Canada has joined the talks, largely to guard against the erosion of its 1989 pact with the United States. This has not appreciably complicated the negotiations. A draft agreement was initialled in August 1992.

Free trade is a highly charged political issue. Much has been said about potential jobs losses and lax workplace safety and environmental enforcement in Mexico. These are serious problems but they have solutions and some are already emerging from the negotiations. Moreover, these concerns distract attention from the more substantial reality – free trade has the potential to initiate, solidify and intensify forces that could substantially transform the Mexican and U.S. economies.

For Mexico, free trade would mark the climax of a radical change in development strategy. In the mid-1980s, Mexico turned away from decades of economic nationalism, which left in its wake crushing debt, an antiquated industrial structure and an institutionalized incapacity for self-sustaining growth. Salinas' reforms are reorienting Mexico to embrace U.S. investment and business culture and to exploit, rather than deny, the mandates of international markets.

For President Salinas, the stakes are high. Rapid labor-force growth, rural overcrowding and increasing demands for social progress require robust growth. Without capital inflows to finance modernization and wider access to foreign markets, his pro-market strategy is severely handicapped. In early 1990, President Salinas visited leaders in Western

---

* This is an early version of the article entitled 'Free Trade with Mexico,' published in *Foreign Policy 87* (Summer 1992) by the Carnegie Endowment for International Peace.

Europe, where he learned that the likelihood of closer commercial ties with the EC were limited. Free trade with the United States is Mexico's best prospect for attracting new capital and expanding exports.

Regarding political reform, Salinas has done more than many observers acknowledge. Most notably, his economic program is disassembling the economic mechanisms that his Institutional Revolutionary Party (PRI) has used to reward supporters and maintain power for six decades. However, a successful transition to stable multiparty government is critically dependent on sustaining growth.

For the United States, a free trade agreement has the potential to be a key element in a new national policy to foster competitiveness. Initially, it will provide U.S. businesses with the opportunity to combine inexpensive Mexican labor with more highly skilled U.S. workers in joint production ventures, much as Japanese competitors are doing in Asia. On another level, though, if coupled with sound worker-adjustment programs, improvements in public education and forward-looking industrial policies, free trade will provide the opportunity to transform the United States into a more knowledge-intensive and wealthier society.

At the same time, free trade would be the best way to support economic and political progress in Mexico and help assure that its new openness and aspirations for partnership with the United States would endure after decades of ambivalence and suspicion about U.S. intentions.

11.2 THE MEXICAN ECONOMY AND REFORMS

Mexico is a large country with much unfulfilled potential. It has a population of about 85 million with literacy rate exceeding 80 per cent. Fifty-seven per cent of its people are under 30, giving it a young, capable and rapidly growing labor force.[1] In 1991, the United States exported $33 billion of goods to Mexico, making it the United States' third largest customer after Canada and Japan.

From the mid-1940s to the early 1970s, Mexico's macroeconomic policy was fiscally conservative. For thirty years, GDP grew at a brisk 6.7 per cent, and inflation averaged only 3.8 per cent.[2]

Like many Latin American countries, though, Mexico sought to industrialize through import-substitution rather than export-led growth. This approach was part of a nexus of foreign investment, industrial, labor, and social policies intended to assert independence from American hegemony. Although import substitution worked well for many years, the oil boom of the 1970s emboldened Mexican officials to pursue nationalist policies with even greater vigor. Restrictions on foreign investment were tightened, and the 1976 Law of Inventions and Trademarks denied pharmaceuticals and other products patent protection. Imports controlled by licenses rose from 65 per cent in 1969 to 100 per cent in 1982.[3] Most commercial banks were nationalized in 1982.

The parastatal sector grew dramatically. By 1975, subsidies to these and other domestic enterprises accounted for 61 per cent of government spending, and the federal deficit was 10 per cent of GDP. By 1982, the year of the first debt crisis, the federal deficit had reached 17 per cent of GDP, as foreign debt swelled to $86 billion and debt service required 34 per cent of export revenues.[4] Although the IMF provided a major loan on the condition that Mexico exercise fiscal restraint, spending soon picked up again. In the end, capital flight, economic contraction and skyrocketing inflation compelled change.

After the 1985 earthquake and the 1986 oil price slide, President Miguel de la Madrid Hurtado (1982–8) initiated and President Salinas accelerated a dramatic change in policy. When inflation soared to 160 per cent in 1987, the business–labor–government Pact for Stability and Economic Growth was launched. Subsidies and spending were slashed. In 1991, the federal deficit was less than 2 per cent,[5] and inflation was 19 per cent – both declined further in 1992.

Structural reform has been breathtaking. President Salinas has imposed the disciplines of competition on Mexican industry. The average tariff has been cut from 29 per cent to about 10 per cent.[6] Import licenses are required for fewer than 5 per cent of products.[7] The range of industries open to foreign ownership has been expanded, and increased foreign participation is permitted in industries such as petrochemicals and insurance.[8] A new intellectual property regime provides a model for other developing countries.

About three-fourths of the 1155 parastatals have been sold, merged or closed.[9] Recent privatizations include the national telephone company, the two national airlines and the four largest banks, as well as holdings in many branches of manufacturing. Also, President Salinas is reforming Mexico's land tenure system. Farmers on *ejidos* – communal estates confiscated under land reform – will now be able to own and sell their plots and enter into joint ventures with foreign investors. This will consolidate small farms, raise productivity and lessen dependence on government credits and subsidies.

Salinas has increased spending on infrastructure and has initiated the National Solidarity Program. Bypassing local politicians and the Mexico City bureaucracy, 'Solidarity' gives money directly to community organizations who propose projects and contribute capital to improve roads, schools, hospitals, sewage and other utilities.[10] Resources go where they are needed most and more progress is achieved per peso – Mexico's traditional overheads of administrative bribes and corruption have been curbed.

Led by maquiladora enterprises,[11] these reforms have attracted job-creating investment back to Mexico. Rapidly increasing exports of manufactures have permitted the GDP to expand at an average rate of 3.7 per cent from 1989 to 1991. However, for Mexico to continue to attract capital, investors must be convinced that economic reforms will continue after Salinas leaves office in 1994. A comprehensive trade agreement with the United States would provide an insurance policy against backsliding and offer good prospects for the continued reduction of statist policies.

11.3 WHAT ARE WE REALLY TALKING ABOUT?

Tariffs do not dominate trade negotiations. Although the United States maintains some high tariffs (duties greater than 15 per cent apply for import-sensitive industries like apparel, footwear and leather products), the average U.S. tariff is less than 3.5 per cent.[12] Apart from the remaining high tariffs, the most significant barriers to Latin American exports to the United States are (1) the arbitrary application of subsidy/countervailing and dumping duties and other unilateral actions under U.S. trade-remedy laws, and (2) management of imports of apparel, steel, sugar, and fruits and vegetables.

For the United States, eliminating Mexican tariffs is important. However, Mexican wages are about one-eighth of U.S. levels,[13] and this offers Mexico much greater advantages in attracting new manufacturing plants than a 10 per cent tariff.

Historically, the real barriers to U.S. sales in Mexico have been import licenses, arcane product standards, discriminatory procurement by government agencies and parastatals, sourcing and production requirements imposed on foreign investors, poor patent protection, sectoral strategies in computers and automotive products, and other methods of import-substitution. Mexico's record in these areas has improved dramatically but more needs to be done. The United States needs a trade agreement that will ensure that Mexico will complete its internal reforms and will afford U.S. products full market access.

Also, lax enforcement of Mexico's fairly stringent workplace safety and environmental rules have attracted smaller enterprises, for example those involved in the furniture industry, away from tougher jurisdictions such as California. This deprives U.S. workers of jobs in much the same way as foreign subsidies do. Moreover, a coalition of U.S. environmental groups, trade unionists and church leaders have made the squalid working conditions in some maquiladoras, and poor air and water quality along the border, hot political issues in Congress. A free trade pact will not pass Congress if it is not accompanied by reasonable assurances that Mexico will enforce its workplace safety laws and initiate a border clean-up plan. This would provide North America with an analog to the EC 1992 Social Dimension.

All of this will require NAFTA to be even more ambitious than the Canada–U.S. FTA, and events are moving in that direction. The structures of NAFTA negotiating groups and the working draft agreement indicate that NAFTA, like the Canada–U.S. FTA, will address tariffs, nontariff barriers, investment rules, trade in services and the application of trade-remedy laws

Mexico has already implemented a policy of only approving new plants if they meet strict workplace safety and environmental standards and it has sought to improve compliance in existing facilities. Since the beginning of 1991, it has shut (generally temporarily) hundreds of plants in the maquiladora region for environmental violations. However, problems persist, and these efforts will take more time, resources and technical expertise to succeed.

In February 1992, the United States and Mexico unveiled a draft environmental clean-up program for the border region that would increase funding for projects and double the number of inspectors in the border area. In addition, negotiators are working on plans to bring Mexican workplace standards up to levels comparable with those required by the U.S. Occupational Health and Safety Administration (OHSA) and to provide technical assistance Mexican inspectors.

Although these moves should substantially address workplace safety and environmental issues, it is unlikely that criticism of free trade from environmental activists and unionists will abate. If NAFTA is approved, these agreements would be only the first of many that gradually moved the United States, Mexico and perhaps Canada toward a common environmental regime. If NAFTA fails, Mexico will have fewer incentives to work closely with the United States on these issues.

The U.S. administration has indicated that a free trade agreement with Mexico is just one element of a broader strategy for the rest of Latin America, and that it wishes to negotiate free trade with other countries that undertake aggressive pro-market reforms. Chile is a prime candidate because it already has a free trade deal with Mexico and its pro-market reforms are the most advanced in the region.

To facilitate accession by Chile and others, NAFTA should be a general agreement as opposed to one focusing on specific trade and investment issues, as is the case between the United States and Mexico. This would avoid a complicated web of bilateral agreements with conflicting commitments. Canada's presence in the negotiations is particularly useful because the Canada–U.S. FTA is the very model of such an agreement. Ensuring that NAFTA is consistent with the Canada–U.S. FTA will help push NAFTA negotiators towards more general provisions in a comprehensive format.

In many ways, a free trade agreement, parallel accords and an accession provision for other Latin American countries, would establish a process that is as broad, though initially not as deep, as the EC agenda; it would create an economic community but without a common external tariff. However, the negotiations required to achieve such integration would extend well beyond the first tariff cuts – fast-track talks would only be a down payment on many years of effort to reconcile the policies and practices of the participating countries. In the end, a common currency and a common external trade policy could become necessary.

Noticeably absent from the NAFTA agenda is illegal immigration from Mexico. This movement of people is largely driven by economic conditions and really cannot be addressed in trade talks. Some NAFTA proponents have argued that free trade, by raising Mexican wages, would diminish the flow of workers. In the longer run, addressing the root causes of poverty by fostering growth, as free trade would do, is the only viable way to eliminate pressures on Mexicans to leave their homeland. In the short run, though, large wage differentials would persist, and free trade, by bringing more Mexicans into urban industrial culture, may actually increase the flow of people across the border.

11.4 OPPORTUNITIES AND ADJUSTMENTS FOR THE UNITED STATES

A North American free trade agreement would result in the more efficient deployment of human resources and capital throughout North America, and would attract more investment from Europe and Asia. By retooling and seeking out more rewarding opportunities, most workers and business would earn higher wages and profits. For the United States, free trade with Mexico would create gains and adjustments that would dramatically exceed those being experienced from the Canada–U.S. FTA owing to three sets of factors.

First, thanks to the 40 years of tariff reductions resulting from GATT, the U.S. and Canadian economies had already been substantially integrated by 1989. Although about 45 per cent of Mexico's exports enter the United States at very low rates of duty through the maquiladora program,[14] the maquiladoras did not become so prominent because U.S. tariffs on other Mexican products were high; rather, they became important because the development of export-oriented production in the traditional Mexican economy was blocked by nationalist policies. These substantially segregated Mexico's traditional industrial sector from the U.S. and Canadian economies, as well as from the maquiladora zone. The most significant effect of a free trade agreement would be not to remove the modest remaining U.S. tariffs. Rather, it would be to reassure Mexican and foreign companies that Mexican reforms are permanent and that it is safe to invest in the modernization of Mexico and to orient new production toward U.S. and Canadian markets.

Similarly, the opening and modernization of an economy the size of Mexico, after so many years of misinvestment and underinvestment, offers U.S. businesses vast new opportunities to sell computers, other sophisticated electrical equipment, industrial machinery, petroleum and mining equipment, and many knowledge-intensive services. The United States enjoys a strong comparative advantage in these areas – for example, exports account for about 45 per cent of U.S. capital-goods production.[15]

In the 1980s, an important impediment to stronger U.S. export performance was the geographic concentration of investment in East and South East Asia, where Japanese firms enjoy marketing advantages through the investments of the *keiretsus* and geographic proximity. An acceleration of Latin American growth and investment, led by Mexico and facilitated by free trade, could instigate a substantial shift in resources to U.S. capital-goods industries and increase the technological-intensity of U.S. manufacturing.

Second, because Mexico has an inexpensive yet literate labor force, free trade would increase the diversity of productive resources available to U.S. businesses in a way that free trade with Canada could not accomplish.

Third, because Mexico's population is so much larger than Canada's and some 20 million live on small agricultural plots where under-employment is a problem, Mexico has many more untapped and underutilized human resources than does Canada. In terms of GDP, Mexico is only about the size of Belgium or the Netherlands but in terms of population and human resources it is about the size of the united Germany.

The integration of the North American economies would fundamentally alter the composition of resources and market opportunities available to U.S. businesses and instigate a major movement of labor and capital from activities emphasizing ordinary factory labor – assembly and simple fabrication jobs – to more technology-intensive pursuits. In turn, the growth of factory jobs in Mexico would offer workers there new opportunities for prosperity.

Although free trade has the potential to be highly lucrative for both the United States and Mexico, this redeployment of resources, being market driven, would impose painful adjustments on many U.S. workers. The most visible manifestation of this would be relentless wage competition from Mexicans for U.S. factory jobs, as well as site-dependent managerial and technical positions, in industries such as apparel, automotive components and assembly, electrical and telecommunications equipment, food products and glass and ceramics.

The United States must accept these adjustments in order to build a more competitive society. For Japan, accessing low-wage factory labor through *keiretsu* investment and trade in East and South East Asia is a key element in its strategy to remain competitive in manufacturing in the face of an appreciating yen and to build a society intensely specialized in high-value, knowledge-intensive activities. If Americans reject free trade with Mexico and protect semiskilled workers from competition, then they must recognize that they are choosing to be a lower-value, lower-income society than Japan.

In many ways the U.S. and Japanese economies are juxtaposed today as the British and German economies were in the 1950s. Germany chose to join Europe, establish educational and worker-adjustment policies responsive to the needs of modernizing industry, and implement industrial policies that fostered the development of leading-edge technologies. Britain chose a nostalgic protectionism for steel and other mature industries. In the end these industries still declined, and Britain was surpassed by Germany in the areas that drove postwar growth – automobiles, electronics and other high technology activities. Today, incomes in Britain substantially lag behind those in Germany. The lesson is clear. If we protect factory jobs today, we consign all our children to a less prosperous future.

The real challenge for proponents of free trade is to recognize the scale, duration and nature of the adjustments and opportunities that free trade would proffer, and to come to terms with the policy responses that will be necessary if the United States is to benefit fully. Three points are important.

First, although free trade would reduce the wage gap between semiskilled workers in the United States and Mexico, it would not do so quickly, and competitive pressures on U.S. jobs would not reduce quickly. In 1990, the wage of the average Mexican industrial worker was 12 per cent of that of his U.S. counterpart. If Mexican real-wage growth were to exceed the U.S. performance by about 7 per cent a year – an heroic assumption – Mexican wages would reach 25 per cent of U.S. levels after 10 years and 50 per cent in about 20 years. Its is interesting to note that in 1980 average wage levels in the four East Asian newly industrializing countries were about 12 per cent of U.S. levels, and in 1990 they had reached about 25 per cent.[16]

Although productivity on East Asian export platforms exceeds 25 per cent of U.S. levels, wages there have not caught up because so many semiskilled workers are flooding the modern Asian economy.[17] The demographics in Mexico and elsewhere in Latin America indicate that much the same can be expected there.

For the time being, low Mexican wages must be considered in juxtaposition to low Mexican productivity; however, much low productivity in Mexico results from the use of outdated capital and poor management in the traditional industrial sector. In the maquiladoras, where firms like Ford and AT&T have invested in modern plants, Mexican productivity is much higher than one-eighth or even one-fourth of U.S. levels. These plants, not facilities further south, are indicative of the kind of competition new Mexican exports will offer American workers.

Second, in industries such as apparel and electronics, moving some assembly and fabrication to Mexico will permit U.S. companies to retain more technical manufacturing jobs, and accompanying design and management positions, in the United States. Economists call this *intra*-industry specialization. For example, the President Warnaco Inc., a Bridgeport, Connecticut-based clothing maker, maintains that 20 per cent of its 8000 U.S. employees are dependent on 1000 workers in Mexico.[18] This story is duplicated elsewhere, and this is an important reason why the normally united and protectionist textile and apparel lobby is split on the issue of free trade.

Although fewer U.S. jobs are lost when U.S. factories relocate in Mexico rather than in Korea or Malaysia, regional effects in the United States can be just as problematic. In 1990, AT&T was able to recall 450 furloughed workers and add 300 new jobs at its Mesquite, Texas electronics plant by moving the work of 1000 employees at its Radford, Virginia plant to Mesquite and to Matamoros in Mexico. On a national basis, only 250 jobs were lost. A far as Radford is concerned 1000 jobs have gone.[19]

Finally, will U.S. firms be able to find enough properly educated and trained workers to provide enough of the knowledge-intensive goods and services that Mexico will need? Evidence is mounting that the answer is 'no'. Many U.S. manufacturers, when confronted with competition from low-wage imports, deskill jobs and increase their reliance on lower-wage, transient labor.[20] They are unable to find sufficient adequately educated and motivated workers to choose technology-intensive options and invest prudently in training.

This is the starkest contrast between the realities of U.S. and Japanese manufacturers. Although U.S. front-line workers are clearly better educated than their Mexican competitors, as the United States and Japan seek to become more knowledge-intensive economies, inadequacies in the abilities of the typical American high school graduate bedevil American employers with problems that Japanese firms just don't seem to encounter.

Additional imports from Mexico would intensify the pressure on employers to deskill jobs and keep wages down. This does not mean that free trade would not raise average U.S. living standards. However, it does mean that free trade would increase the incomes of the well-educated and highly skilled at the expense of workers with only general high school backgrounds and little other training. Free trade would exacerbate the trend towards a less equal distribution of income. This dynamic may be one of only several factors pushing labor markets in this direction, but it would have unwelcomed consequences for maintaining political support for free trade over the ten to twenty years necessary to fully implement NAFTA.

To ease worker adjustments and maintain political support, free trade with Mexico should be seen as one element of a multifaceted national policy to improve competitiveness. Tariffs and other trade barriers must be phased out at a pace that will permit retirements and normal retirements to absorb as many of the job losses as practical. In the Canada–U.S. FTA up to 10 years is provided. In NAFTA, adjustments will be more difficult; therefore a 15-year transition period for apparel and several other sensitive sectors seems appropriate.

School reforms will be required. The U.S. administration has advocated the introduction of national standards for public schools that are responsive to the requirements of technologically sophisticated workplaces. However, the vast majority of the workers who will be in the labor force of 10 or 15 years hence will have already completed their education. Therefore a national retraining program should be set up to help front-line workers reclaim lost skills and acquire new ones.

It is often difficult to determine whether workers have lost their jobs because of competition from imports or from other factors, such as technological change. Retraining assistance should be made available to all permanently laid-off workers, regardless of cause. To help even out regional imbalances in adjustment costs, retraining assistance should be made available to all workers, employed or not, in communities designated as economically distressed.

Although the United States has a strong position in capital goods and related services, in recent years it has lagged behind Japan in R&D spending and new investment. Urgently needed are a national development bank to finance export-oriented investments in rapidly expanding industries and a civilian analogue to the Defense Advanced Projects Research Agency to assist commercially promising, precompetitive research.

In the end, the United States must recognize the need to combine good international policy with good domestic policy. It will not reap the benefits of free trade if its workers are not equipped with the skills and tools necessary to exploit the opportunities offered by NAFTA.

11.5 ECONOMIC AND POLITICAL ADJUSTMENT IN MEXICO

In Mexico, free trade would impose even more difficult adjustments. Although free trade would expand markets for activities requiring inexpensive factory labor, the plants and firms poised to exploit these opportunities often are not the ones that benefited from import-substitution. For many manufacturers of products such as automotive components, textiles and industrial machinery, greater *intra*-industry trade and specialization will require modernization, downsizing, and sometimes, plant closures. This will mean difficult times for many workers in the traditional industrial centers of Mexico City, Monterrey and Guadalajara.

Many new jobs will be provided by U.S. and other foreign firms, as well as by reorganised Mexican suppliers. After Pemex, Ford and General Motors are Mexico's largest exporters.[21] Also important are IBM, DuPont, Celanese, Motorola, Hewlett Packard, Nissan and Volkswagen. However, Mexico has strong enterprises of its own that will penetrate markets throughout North America, such as Vitro in glass products, Cemex in cement, Altos-Horn in steel, Petrocel in petrochemicals and the Alpha group in packaging, petrochemicals and autoparts.

In a textbook example of intra-industry specialization, Vitro and Corning are combining their consumer housewares divisions. By accessing Corning's U.S. and global distribution networks, Vitro will aggressively expand sales of high-quality glass, crystal and ceramic tableware, while its U.S. partner should be able concentrate more on its high-technology activities.

Although President Salinas has placed economic progress ahead of political reform, he has contributed importantly to cleaning up Mexican politics. He has established a voter identification system to combat election fraud and has created public expectation of honest elections. Following the August 1991 mid-term elections, two suppo-sedly victorious PRI candidates were confronted with allegations of election fraud; following civil unrest over the matter Salinas pressured them in to stepping down. Although the PRI won overwhelming majorities in other elections, confirming the popularity of Salinas' economic programs, these incidents indicate that the PRI must bend to calls for honest elections or resort to self-destructive repression.

To fully understand the political significance of Salinas, though, it is necessary to appreciate how the PRI has maintained its supremacy for more than 60 years. It is organized into three constituencies – the popular (middle-class professionals, including government workers), labor and farm sectors. Under the old regime, leaders in each sector enjoyed access to political power and economic benefits, which they used to control and broker support for the PRI among their constituents. For example, the petroleum workers union awarded some Pemex contracts, creating profitable opportunities for its leaders and the resources to reward its members. Political allegiance was easily obtained from farmers through government control of land tenure on *ejidos* and farmers' dependence on federal credits and other assistance. By phasing out most price controls, import licenses and other mechanisms of a state-managed economy, economic reforms are circumscribing the power of bureaucrats and the opportunities that were open to party stalwarts in government service. Planned education reform will erode the PRI's strength with teachers. The labor sector was dealt a significant blow when Salinas arrested the head of the petroleum workers union and reformed Pemex procurement procedures. More broadly, the elimination of price controls, subsidies and protection means that unions are less able to deliver government favors that insulate workers from competitive realities. Agricultural reform will emancipate *ejidatarios* from a neofuedal system of control.

In the end, the PRI will have to accede to genuine multiparty competition and share more political power with the opposition Nation Action Party (PAN) and the Democratic Revolutionary Party (PDR), which is led by leftist Cuauhtemoc Cardenas Solorzano. Does this mean that Salinas, like Gorbachev, has unleashed forces that will lead to his own demise and that of the PRI?

Hardly. Salinas has avoided Gorbachev's mistake of dissolving the central authority before economic reforms have created the transformations necessary to make them selfsustaining or the prosperity to support stable multiparty democracy. Moreover, President Salinas seems to be preparing the PRI for the day when erosion of the old levers of control will require its candidates to win elections through grassroots campaigning. He has fostered more open decisionmaking inside the PRI and the party is offering more attractive candidates – younger men and women with local political bases are replacing party hacks. Similarly, through Solidarity, Salinas has built new bases of support for PRI candidates within community organizations, and several prominent Solidarity figures won senate seats in 1991.

In all of this, free trade would strengthen support for economic reform by increasing investment and technology transfers, thereby raising the incomes of businesses, professional and ordinary workers and farmers making products for sale in the United States in a rationalized Mexican economy. In turn, this would provoke additional economic reforms, hasten the erosion of the PRI's economic levers of political control, and contribute to the prosperity needed to sustain multiparty democracy. Moreover, an agreement with the United States would make it very difficult for future Mexican governments to offer political operatives new opportunities for control by backsliding on reforms without violating the trade pact.

If free trade does not materialize – because the U.S. Congress decides not ratify an agreement – Mexico's ability to attract capital and increase exports at the pace necessary to sustain economic reform would be severely handicapped.

A U.S. rejection of Salinas' free trade initiative would give his critics on the left an opening. Although it is doubtful whether the PDR or the PAN could achieve victory in a presidential election, a failure to achieve free trade would wound Salinas and affect the policies of the next president. It would be difficult to reverse those economic reforms already in place; however a more left-leaning or less pro-reform administration could slow or halt the process. Some protectionist measures could be reintroduced, and administrative corruption could increase its drain on private-sector efficiency, reawaken hyperinflation and sabotage the confidence of domestic and foreign investors.

11.6 SHAPING THE AMERICAN RESPONSE

The opening up of Mexico will necessitate a bold American policy. Serving U.S. interests will require thorough assessment and decisive action, not timidity. Mexico and the rest of Latin America offer prospects for vast new markets and the opportunity for Americans to build a more knowledge-intensive and much wealthier society.

Legitimate concerns have been raised about environmental and workplace safety standards in Mexico. However, throughout the negotiations Mexico has demonstrated its desire to bring its standards up to the levels of industrialized countries. The proposed border cleanup pact, and the workplace safety accord that will follow, offer the United States a critical opportunity that will be lost if free trade languishes.

Worker adjustments can be handled best by extending the elimination of tariffs and quotas. Marrying such an approach with improvements in education and training programs and with industrial policies that encourage export-oriented industries would make it possible to create high-wage jobs and exploit U.S. technological advantages in capital goods and knowledge-intensive services.

Japan is jettisoning low-skill factory jobs and creating a knowledge-based economy while the *keiretsu* manage its expansion of trade and investment in Asia. If the United States does not respond by similarly combining its energies with Mexico, it cannot expect to compete with, nor to have as high a standard of living, as Japan.

For decades Americans have been nagging and cajoling Latin Americans to open their markets to U.S. goods and investment and let the compelling energies of market capitalism and entrepreneurship transform their societies. More than any other figure, President Salinas personifies the new pro-market ethos that is sweeping Latin America and disassembling the bulwarks of statism and corporatism. The United States' response to his request for free trade will be a defining moment for U.S. relations with the entire region.

To date, President Salinas' ability to act decisively and pay minimal deference to corporatist vested interests has distinguished Mexican reforms from most other Latin American experiences. Free trade would provide Salinas with the opportunity to solidify these gains and place Mexico firmly on a non-statist free market path and to pursue political reform.

U.S. rejection of Mexico's offer to enter into a free trade pact would undermine Salinas, and any hint that a more left-leaning PRI government might reassert even minimal elements of discarded statist policies could frighten foreign investors and jeopardize Mexico's recovery.

The United States must recognize both the opportunities that reform in Mexico offers and that a failure to embrace free trade would adversely affect economic and political progress there. The status quo is not an option, because events are moving so quickly and recovery is so fragile in Mexico that there really is no status quo in Mexico, nor in U.S.–Mexican relations.

NOTES AND REFERENCES

1. Committee for the Promotion of Investment in Mexico, *Mexico: Economic and Business Overview* (Mexico City: June 1990).
2. John H. Purcell and Dirk Damrau, *Mexico: A World Class Economy in the 1990s* (New York: Salomon Brothers Sovereign Assessment Group, 1990) p. 2.
3. U.S. International Trade Commission Report 2275, April 1990, p. 4.5.
4. Ibid., pp. 1.2, 1.3.
5. U.S. Embassy, Mexico, *Current Economic Trends Report*, August 1991.

6. U.S. International Trade Commission Report 2353, February 1991, p. 1.2.
7. Department of State, 'Mexico's Economic Reforms', mimeo, 1991.
8. See Peter Morici, *Trade Talks with Mexico: A Time for Realism* (Washington, DC: National Planning Association, 1991) pp. 21–3, 27–9.
9. U.S. Embassy Mexico, *Current Economic Trends Report*, August 1991.
10. George W. Grayson, 'Mexico's New Politics', *Commonweal*, 25 October 1991, pp. 612–14.
11. Maquiladora factories assemble U.S. components; duties are assessed only on Mexican value added when products reenter the United States.
12. U.S. International Trade Commission Report 2353, February 1991, p. 2.2.
13. The U.S. Bureau of Labor Statistics estimates that Mexican compensation in manufacturing was 12 per cent of U.S. levels in 1990; see U.S. Department of Labor, BLS Report 817, November 1991.
14. U.S. International Trade Commission Publication 2353, February 1991, p. 1.6.
15. Lawrence B. Lindsey, 'America's Growing Economic Lead', *The Wall Street Journal*, 7 February 1992, p. A14.
16. U.S. Department of Labor, BLS Report 817, November 1991.
17. The EC enjoyed much success by absorbing Portugal and Spain. However, it is important to recognize that Mexico's population is 85 million – about 30 per cent of the United States and Canada, while the population of Portugal and Spain are 50 million – about 15 per cent of the EC. Also, at the time they joined the EC, Portugal and Spain had much higher wages than Mexico does now, and they became eligible for community-wide regional development programs.
18. Stuart Auerbach, 'Splitting Protectionist Seams: Mexican Trade Pact Unravels the Once-Durable Textile Lobby', *The Washington Post*, 12 May 1991, pp. H1, H8.
19. Frank Swoboda and Martha M. Hamilton, 'How Virginia Lost Jobs to Texas, Mexico: Closing of AT&T Plant Suggests Complexity of Free Trade Issue', *The Washington Post*, 5 May 1991, pp. H1, H6.
20. See Commission on the Skills of the American Workforce, *America's Choice: High Skills or Low Wages!* (Rochester, NY: National Center on Education and the Economy, 1990).
21. *America Economica* (New York: Dow Jones, December 1991) p. 59.

12 An Empirical Estimation of the Level of Intra-Industry Trade between Mexico and the United States*

Jorge G. Gonzalez and Alejandro Velez

12.1 INTRODUCTION

This chapter presents an evaluation of the level of intra-industry trade between Mexico and the United States. The calculated indexes of intra-industry trade indicate a rapid increase in this type of trade during the 1982–90 period. Additionally, the current level of intra-industry trade between the two nations is quite high when compared with similar indexes of other nations. These results help to explain the apparent ease with which the United States adjusted to increased Mexican imports during the 1980s. Furthermore, the high level of intra-industry trade indicates that after NAFTA has been implemented there should be no major dislocation of productive activities in either of these countries as a result of the expansion in trade.

In 1992 there was widespread debate in the United States about the possible consequences of a North American Free Trade Agreement (NAFTA). Opinions on this issue vary widely. Some predict catastrophic effects while others foresee tremendous benefits. Although both views are probably exaggerated, the negative one seems more so, especially when talking about U.S.–Mexico trade. If one looks at the growth of U.S. imports from Mexico over the past few years, one discovers a dramatic increase. From 1982–90 total U.S. imports increased by 93 per cent, or at an annual rate of 8.4 per cent.[†] In

* The authors wish to acknowledge gratefully the research and computation assistance provided by Lourdes Hernández. A previous version of this chapter was presented at the November 1991 Southern Economic Association meetings in Nashville.
† Total imports went from $15 770 million to $30 172 million.

spite of this there were no dramatic changes in U.S. employment nor a major displacement of workers. This growth in trade was accompanied by major structural change. Total imports, excluding petroleum increased by 214 per cent over the same period, or at a annual rate of 15.4 per cent. Thus the composition of Mexican exports shifted away from oil and towards manufactured goods, which could potentially be in competition with domestically produced ones. As a result of all of these changes, Mexico has already become the United States' third largest trading partner, even without a free trade area.

In order to understand why, in the face of rapidly increasing trade, U.S. import-competing industries have not faced major employment changes as some critics would expect, one must look at the specific characteristics of this increased trade between the two countries.

One possible explanation of the apparent ease of adjustment lies in the distinction between inter-industry and intra-industry trade. It has been argued in the literature that adjustment to increased intra-industry trade is easier than that to inter-industry trade,[1] that if intra-industry trade expands, then workers in import-competing industries have to shift to the production of specific lines of output within the same industry. This adjustment is easier to accomplish than the shifting to different industries that accompanies increases in inter-industry trade. In the context of Mexico–U.S. trade, high intra-industry trade would help explain the ease of adjustment to increased trade in the 1980s. And even more importantly, it would signal that after NAFTA has been approved and the countries expand their trade further, future adjustment costs would be much smaller than alarmists have been predicting.

The purpose of this chapter is to measure the level of intra-industry trade between Mexico and the United States and to evaluate its pattern for the 1982–90 period. The chapter is organized as follows. Section 12.2 defines intra-industry trade and discusses the indicator used in this study to measure intra-industry trade. Section 12.3 presents empirical measurements of the intra-industry trade between the two countries and evaluates their changes. Finally, the conclusions drawn and the agenda for further research are discussed.

12.2   THE MEASUREMENT AND INTERPRETATION OF INTRA-INDUSTRY TRADE

Intra-industry trade has been defined as the coexistence of imports and exports within the same industry in a given period of time. Researchers have been interested in this type of trade not only because of its distinct adjustment characteristics but also because it seems to defy specialization by industries or sectors as predicted by the Heckscher–Ohlin model.

Several variables have been identified in the literature as being determinants of the level of intra-industry trade among countries. A high domestic income, an advanced level of economic development and similarity of income levels among trading partners tend to produce high volumes of intra-industry trade. Geographic proximity, foreign direct investment and a high level of economic integration also tend to facilitate this type of trade. At the industry level, industries with important economies of scale and/or product differentiation are expected to present high levels of intra-industry trade.

The literature of intra-industry trade presents several possible indicators to measure the importance of this type of trade.[2] The most widely used measurement is the one proposed by Grubel and Lloyd (1971, 1975). Their basic index measures the intra-industry trade in industry i for country j as:

$$B_{ij} = \frac{(X_{ij} + M_{ij}) - |X_{ij} - M_{ij}|}{(X_{ij} + M_{ij})} \cdot 100 \tag{12.1}$$

where X_{ij} and M_{ij} represent country j's exports and imports of commodity i.

The intra-industry trade index, B_{ij}, measures intra-industry trade as a proportion of total trade in the industry.[*] This index ranges in value from zero to 100. A zero value signifies that all trade in the industry is of the inter-industry type. As the proportion of trade represented by intra-industry trade increases, the value of B_{ij} approaches 100.[†]

In order to find an aggregate index for the level of intra-industry trade of country j, one can compute a weighted average of the B_{ij}s.[§]

---

[*] The difference between exports and imports in a given industry, or $X_{ij} - M_{ij}$, can be interpreted as inter-industry trade. Therefore, the numerator of equation (12.1) measures intra-industry trade by subtracting inter-industry trade from total trade.

[†] Equation (12.1) can be rewritten for convenience as follows:

$$B_{ij} = \left[1 - \frac{|X_{ij} - M_{ij}|}{(X_{ij} + M_{ij})} \right] \cdot 100 \tag{12.3b}$$

[§] The relative size of exports plus imports of the industry with respect to total exports and imports is used as the weights for the computation of B_j. Drabek and Greenaway (1984) criticize the use of the weighted average, and propose instead the use of simple unweighted averages. However, most studies use weighted averages.

This index, B_j, can be expressed as follows:

$$B_j = \frac{\sum_{i=1}^{n}(X_{ij} + M_{ij}) - \sum_{i=1}^{n}|X_{ij} - M_{ij}|}{\sum_{i=1}^{n}(X_{ij} + M_{ij})} \qquad (12.2)$$

where n represents the total number of industries in the country.
Equation (12.2) can be rewritten as:

$$B_j = \left[1 - \frac{\sum_{i=1}^{n}(X_{ij} - M_{ij})}{\sum_{i=1}^{n}(X_{ij} + M_{ij})}\right].100 \qquad (12.3)$$

Two complications have been associated with the computation of B_j as a measurement of intra-industry trade. One deals with the bias introduced to the index when there is an overall trade imbalance and the other deals with problems associated with the aggregation of different industries within one 'industry' classification. The trade imbalance and the categorical aggregation problems have been discussed at length in the literature and several possible adjustments to the index have been proposed.[3] However at this point there is no consensus as to how to correct these problems or even as to whether these problems should be corrected at all. This discussion is beyond the scope of this chapter. In order to facilitate comparisons between its results and those found by other authors, the unadjusted B_j is used throughout the chapter.

Most empirical studies of intra-industry trade have been conducted at the 3-digit level of the SITC classification. The 3-digit level has been preferred because it closely resembles the definition of an industry. Furthermore, at this level of aggregation comparable data is available for a large number of countries.

12.3 INTRA-INDUSTRY TRADE BETWEEN MEXICO AND THE UNITED STATES

Existing literature does not provide calculations of the level of intra-industry trade between Mexico and the United States. However, a few multi-country studies provide estimates of the level of intra-industry trade between Mexico and the rest of the world. Aquino (1978), using an industry classification different from the one used here, finds that Mexico's intra-industry trade in the manufacturing sector was 36.6 per cent of total sectoral trade in 1972.[4] Using 1974 data and yet another industry classification, Balassa (1979) computes four different intra-industry indexes for Mexico's manufacturing sector. He finds indexes of 34.3, 24.6, 46.3, and 36.2 respectively for trade between Mexico and the rest of the world, developed countries, LAFTA countries and LDCs that are not LAFTA members. Finally, Havrylyshyn and Civan (1983) compute the intra-industry trade index for Mexico's manufacturing sector with 1978 data. They find this index to be equal to 31.9. Although none of these calculations are directly comparable with the ones to be presented here, a definite pattern seems to emerge. It appears that the index of intra-industry trade for Mexico's manufacturing sector during the 1970s fluctuated in the 30s range. These figures are consistent with the figure found in this study for the first year of analysis (see Table 12.1).

Consistent with previous empirical work for other countries, this study uses data at the 3-digit and 4-digit levels of the SITC to compute the intra-industry trade index for commerce between Mexico and the United States. The data includes all industries at the respective SITC levels at which recorded trade took place.[5] The discussion covers the period from 1982 to 1990.*

Table 12.1 presents the results of the computations of the intra-industry index as given by equation (12.3) at the 3-digit SITC level. In this table, the weighted average of each of the ten 1-digit SITC categories are shown for each year of the study period.[†] Then the intra-industry index for the whole economy is given. Since most studies eliminate SITC category 9 from the estimation, the index without this category is also reported.[§] Finally, it is expected that intra-industry trade should prevail in the manufacturing sector. Because of this fact, several studies have concentrated on this sector only. This manufacturing-only index is also reported in Table 12.1.[‡]

The intra-industry indexes presented in Table 12.1 show a consistent increase across the board. All industry groups, with the exception of

---

* This period was chosen because it was during this time that trade between Mexico and the United States underwent very rapid growth. It is also the most suitable bench mark for discussing developments in the 1990s under NAFTA.
† Each of the indexes is computed the same way as B_j in Equation (12.3). However, for this case, n represents the number of 3-digit industries within the one-digit category and not the total number of industries in the country.
§ Category 9 of the SITC is the 'others' section. Examples of items in this category are zoo animals, military goods, gold, metal coins, and so forth.
‡ The SITC categories 5, 6, 7 and 8 are considered to represent the manufacturing industry.

Table 12.1   Intra-industry trade between Mexico and the United States
(3-digit calculation)

SITC section	1982	1983	1984	1985	1986	1987	1988	1989	1990
0 Food and live animals	19.4	5.2	9.5	17.3	14.5	9.3	17.9	18.0	22.8
1 Beverages and tobacco	3.2	1.4	3.7	2.1	2.0	3.3	10.0	28.3	16.3
2 Crude materials, inedible, except fuels	26.5	27.3	18.7	20.0	24.2	28.6	27.1	39.4	42.7
3 Mineral fuels, lubricants and related materials	8.5	5.5	8.5	7.3	12.6	15.4	18.8	11.3	14.7
4 Animal and vegetable oils, fats and waxes	0.5	0.0	1.8	1.1	0.5	5.8	10.7	22.5	22.7
5 Chemicals and related products	9.1	15.2	25.7	26.9	26.2	25.2	28.1	39.6	40.5
6 Manufactured goods classified chiefly by material	40.5	47.0	50.0	51.4	56.8	58.7	62.7	59.6	54.3
7 Machinery and transport equipment	43.7	53.1	56.2	49.9	51.1	51.8	52.0	68.6	65.3
8 Miscellaneous manufactured articles	37.8	46.8	47.0	43.2	46.5	44.4	49.4	70.0	71.4
9 Commodities and transact. not class. elsewhere	42.9	43.2	53.1	18.9	32.2	29.3	28.1	33.2	15.4
All SITC sections (0–9)	25.9	26.1	33.2	31.7	38.4	40.1	43.5	53.0	51.1
All SITC sections except 9 (0–8)	25.4	25.8	32.6	32.1	38.6	40.6	44.1	54.1	53.0
Manufacturing industry (5–8)	38.6	46.8	50.6	46.9	49.1	49.8	51.3	65.3	62.8

category 9, show increases in their level of intra-industry trade. As a result the overall indexes also show consistent increases from 1982 until 1990. Furthermore, as has been the case in studies of intra-industry trade among developed countries, intra-industry trade is more prevalent in the manufacturing sector.

When one studies the trade between Mexico and the United States, one must be aware of the relative importance of Mexican oil exports to the United States. During the period of this study the participation of oil exports in total Mexican exports have ranged from almost 50 per cent to less than 15 per cent. It is clear that these variations in oil exports, not only in relative but also in absolute terms, have a strong impact on the computed indexes of intra-industry trade. Furthermore, since these changes are due to variations in the world price of oil and have little to do with the structure of bilateral trade, it makes sense to recalculate the intra-industry trade indexes leaving out the exports of crude petroleum. The results of this adjustment are presented in Table 12.2.

Table 12.2   Intra-industry trade between Mexico and the United States
(3-digit calculation without crude petroleum)

SITC section	1982	1983	1984	1985	1986	1987	1988	1989	1990
3 Mineral fuels, lubricants and related materials (excl. crude petroleum)	41.7	37.4	44.7	43.4	60.0	79.5	85.2	56.5	68.9
All SITC sections (0–9)	36.0	36.8	42.9	40.6	43.2	44.7	46.5	57.4	55.7
All SITC sections except 9 (0–8)	35.7	36.6	42.6	41.4	43.6	45.4	47.4	58.9	58.1
Manufacturing industry (5–8)	38.6	46.8	50.6	46.9	49.1	49.8	51.3	65.3	62.8

Table 12.2 presents first the adjusted intra-industry trade index for category 3 of the SITC, which includes mineral fuels, lubricants and related materials. Then the adjusted values of the overall indexes calculated in Table 12.1 are shown. The overall results of the adjustment show an increase in the calculated indexes of intra-industry trade. This result derives from the fact that a large portion of inter-industry trade has been discarded from the calculations. Furthermore, the relative increase in the indexes is more pronounced in the early years of the study period than in the latter years. This is due to the relative decline in the importance of oil in the overall trade picture between the two countries. In summary, when one ignores trade in crude petroleum, the level of intra-industry trade between Mexico and the United States becomes higher; however its rate of increase between 1982 and 1990 is somewhat smaller.

As mentioned before, categorical aggregation is one of the problems that arise when estimating intra-industry trade. Some authors have argued that there is enough factor variability within the 3-digit SITC industries so that the 4-digit level would provide a more suitable definition of an 'industry'. One would expect that as the level of aggregation is reduced, the level of intra-industry trade should fall.

However, if the measured intra-industry trade at the 3-digit level is not merely a result of misclassification problems, then the 4-digit levels should be close to the 3-digit indexes and behave in a similar way. The calculated intra-industry trade indexes at the 4-digit level are presented in Tables 12.3 and 12.4. The results are consistent with a priori expectations. As the level of aggregation is reduced, the level of intra-industry trade falls. Nevertheless, the measured levels of intra-industry trade are still substantial and also show a rapid increase. When one compares the overall indexes at the 3- and 4-digit levels, the proportional difference between the two has declined over time. This result could be interpreted as a sign that the categorical aggregation problem of the 3-digit indexes has declined over time for the trade between Mexico and the United States.

When one compares the levels of intra-industry trade found here to the levels found in the literature for other countries, one finds that intra-industry trade between Mexico and the United States approaches the levels that exist between developed countries. Furthermore, this level is much higher than the intra-industry trade that exists between most developed and developing nations.[6] These findings appear to defy theoretical predictions. The vast differences between the economies of these two countries make them poor candidates for high intra-industry trade.

One can speculate as to the reasons behind the unusually high indexes of intra-industry trade between Mexico and the United States. Of the determinants found in the literature, clearly geographical proximity and investment flows are of particular importance. These two countries have the largest border between a developed and a developing country. Furthermore, the transportation costs between the two countries are much lower than those with their trading partners outside North America, even after one takes into account the difficulties and additional costs imposed by the poor Mexico–U.S. border infrastructure. Likewise, U.S. companies have made substantial investments in Mexico. The original maquiladora program and, more recently, the liberalization of foreign investment regulations in Mexico have attracted considerable U.S. investment funds into Mexico's manufacturing sector. Several U.S. industries have made manufacturing processes in Mexico an integral part of their global production.* As

---

* The automobile and, to a lesser degree, the computer industries are examples of this integration. See Weintraub *et al.* (1991) for an evaluation of the economic integration between Mexico and the United States.

Table 12.3   Intra-industry trade between Mexico and the United States
(4-digit calculation)

SITC section	1982	1983	1984	1985	1986	1987	1988	1989	1990
0 Food and live animals	5.1	2.9	5.2	12.1	8.0	6.1	13.0	10.3	10.4
1 Beverages and tobacco	3.2	0.3	0.7	2.0	1.0	2.4	8.8	23.3	13.0
2 Raw materials, inedible, except fuels	10.5	13.9	10.1	10.6	15.8	19.9	18.4	22.2	23.5
3 Mineral fuels, lubricants and related materials	5.9	4.8	8.0	6.1	11.3	9.9	12.1	6.4	9.3
4 Animal and vegetable oils, fats and waxes	0.3	0.0	1.2	0.0	0.0	0.0	0.0	19.8	14.2
5 Chemicals and related products	6.0	7.5	9.4	10.8	14.3	13.9	16.0	27.7	30.9
6 Manufactured goods classified chiefly by material	17.9	21.7	22.1	21.1	23.7	27.3	32.3	46.1	41.3
7 Machinery and transport equipment	32.0	40.0	42.0	36.9	38.3	38.0	41.5	56.6	53.3
8 Miscellaneous manufactured articles	28.5	37.5	39.8	36.0	38.2	36.8	41.6	62.9	64.4
9 Commodities and transact. not class. elsewhere	27.5	25.4	37.9	1.7	18.2	19.8	17.7	33.2	15.4
All SITC sections (0–9)	16.8	18.1	23.2	21.4	26.7	27.6	31.9	42.9	40.7
All SITC sections except 9 (0–8)	16.5	18.0	22.8	22.0	27.0	27.9	32.5	43.4	42.1
Manufacturing industry (5–8)	26.6	32.6	34.9	31.8	34.1	34.4	38.1	53.7	51.6

Table 12.4   Intra-industry trade between Mexico and the United States
(4-digit calculation without crude petroleum)

SITC section	1982	1983	1984	1985	1986	1987	1988	1989	1990
3 Mineral fuels, lubricants and related materials (excl. crude petroleum)	28.8	32.3	41.9	36.0	53.9	51.2	55.0	32.2	43.6
All SITC sections (0–9)	23.4	25.6	30.0	27.4	30.0	30.7	34.1	46.4	44.4
All SITC sections except 9 (0–8)	23.2	25.6	29.8	28.3	30.5	31.2	34.9	47.2	46.1
Manufacturing industry (5–8)	26.6	32.6	34.9	31.8	34.1	34.4	38.1	53.7	51.6

a result of these investment flows, the economies of Mexico and the United States have become highly integrated in certain sectors. This trend, which clearly accelerated in the 1980s, is a major factor behind the high level of intra-industry trade found between the two countries.* Agmon (1979) notes that in the presence of intra-industry trade the standard result of trade theory, that capital mobility is a substitute for trade in goods, does not apply. With intra-industry trade, capital mobility and international trade become complements. The case of Mexico and the United States appears to give further support to this hypothesis of complementarity. This complementarity can be explained by the trade in intermediate and final goods between the firm's headquarters and its foreign operation. The higher the activity abroad, the larger the amount of trade that will occur within the firm, and therefore within the industry.[7] This is clearly what occurs with the U.S. maquiladoras located in Mexico.

The industrial policies of Mexico and the United States have reinforced this complementarity between investment and trade. According to Weintraub (1991, p. 51) the import substitution strategy followed by Mexico after the Second World War unintentionally 'encouraged complex linkages with foreign industries . . . these affiliations were predominantly with U.S. companies'. For example, Mexico's stringent domestic content rules forced U.S. companies to form many partnerships with Mexican investors. Thus, while Mexico pursued greater independence from the United States, it actually forged alliances between affiliates which eventually resulted in a high incidence of simultaneous purchases and sales within the same industries across borders. Furthermore, low U.S. nominal and effective tariffs on intermediate products, for example automobile parts, have reinforced this process.

One can further speculate about the possible effects of NAFTA by comparing the current level of intra-industry trade between Mexico and the United States with that of the European countries before they started their economic integration. Grubel and Lloyd (1975) calculate intra-industry indexes for the six original members of the EC. The intra-industry index for all trade between the United States and Mexico

---

* This pattern of high intra-industry trade was also encountered by Culem and Lundberg (1986) in the trade between Japan and the Asian NICs. Geographic proximity and foreign investment there also stimulated high economic integration with the resulting high level of intra-industry trade. Nevertheless, the intra-industry trade between Mexico and the United States is much higher than that found for Japan–Asian NICs by these authors.

for 1990 is higher than that for Italy, France and West Germany in 1959, and only slightly lower than that of Belgium–Luxembourg and the Netherlands. If the impact of crude petroleum is eliminated, then the index for U.S.–Mexico trade is higher than that of all EC countries when they formally started their economic integration.*

It has been widely documented in the literature that the formation of the EC contributed to a rapid increase in the level of intra-industry trade among its members.[8] The apparent ease of adjustment to the expansion in trade in Europe seems to derive from the fact that a large portion of Europe's trade was intra-industry and not inter-industry. It appears that given the already high degree of intra-industry trade shared by Mexico and the United States, the adjustment process in these countries after NAFTA has been enacted should follow a similar path to that followed by Europe. In other words, the high level of intra-industry trade between the two countries indicates that NAFTA should not create major adjustment problems in these countries.

12.4 CONCLUSIONS

The findings of this chapter indicate that the level of intra-industry trade between Mexico and the United States increased substantially from 1982–90. In addition, the current level of intra-industry trade between these nations is quite high when compared with similar indexes of other nations. These results help to explain the apparent ease of adjustment to increased Mexican exports to the United States during the 1980s. Furthermore, given the high level of intra-industry trade that is already taking place, one can expect that the increased trade which would take place after the implementation of NAFTA would not have a major disruptive effect on the economy of either country. The catastrophic predictions of some observers are founded on a misunderstanding of the structure of trade between these countries. In the future one can expect the level of intra-industry trade to increase still further.

The study of the structure of Mexico–U.S. intra-industry trade has been neglected in the literature. The research presented in this chapter is qualitatively different, and thus, should contribute to the growing

---

* The 3-digit calculations presented in Tables 12.1 and 12.2 are the ones used for this comparison. This is done because Grubel and Lloyd (1975) use the 3-digit classification in their study.

literature on the trade between Mexico and the United States. This chapter has concentrated on evaluating the level of intra-industry trade. However, more work needs to be done. The determinants of intra-industry trade should be evaluated further, both at the national level and at the industry-specific level. It was mentioned above that the presence of maquiladoras has been a promoting factor for intra-industry trade. Subsequent studies should isolate this sector and also study factor intensity, product differentiation and FDI in the maquila industry. Canada should also be included in future studies since it will also be a member of NAFTA.

The economic consequences of a NAFTA can only be understood if all aspects of the trade between the North American countries are studied. The existing high levels of intra-industry trade bode well for NAFTA. However, further studies and estimates are needed.

NOTES AND REFERENCES

1. Among others, Adler (1970), Aquino (1978), Balassa (1965), Caves (1981), Cox and Harris (1985), Hufbauer and Chilas (1974) and Krugman (1981) have argued this point in the literature.
2. Different measurements are proposed by Verdoorn (1960), Balassa (1966), Grubel and Lloyd (1971, 1975), Aquino (1978) and Bergstrand (1983).
3. See Grubel and Lloyd (1975), Aquino (1978, 1981), Greenaway and Milner (1981, 1987) and Bergstrand (1983), among others, for further discussion on the trade imbalance effect. Categorical aggregation has been discussed, among others, by Finger (1975), Gray (1979) and Greenaway and Milner (1983, 1987).
4. Aquino (1978) computes several different indexes of intra-industry trade. The index reported here is the one comparable to the indicator used in this chapter.
5. The sources of the data are the U.S. Department of Commerce, Bureau of the Census publications *U.S. General Imports* and *U.S. Exports of Domestic Merchandise*, and the data tapes of the Foreign Trade Division of that bureau.
6. See Culem and Lundberg (1986) for a recent study comparing intra-industry trade between developed economies, and between developed and developing nations. In a similar result Glejser *et al.* (1982) find high intra-industry trade among 'high-wage' countries, and low for 'low-wage' nations.
7. Kol and Rayment (1989) study trade of intermediate goods as a form of intra-industry trade.
8. Among others, Balassa (1966), Grubel (1967) and Greenaway (1987) have found increases in intra-industry trade in European countries after the formation of the EC.

13 U.S. and Mexican Foreign Exchange Risk Management Techniques

Kurt R. Jesswein, Stephen B. Salter and
L. Murphy Smith

13.1 INTRODUCTION

This chapter will compare the usage and perceptions of a selection of foreign exchange risk management products among large U.S. and Mexican companies. Some significant differences in the utilization of products, perspectives on key factors in product selection, and attitudes to innovation can be found. The chapter will also explore factors that may influence these differences.

In many respects foreign exchange risk management (FERM) and its direct sibling, financial product innovation, is changing the way financial markets operate. Yet much of the research has concentrated on descriptions of the products on offer, or environmental influences leading to the use of products, without placing the use of particular products in a specific geographic, legal or cultural environment. This runs contrary to the existing literature in the finance area, which shows that substantial segmentation still exists in global capital markets (summarized in Salter, 1991, pp. 1–4). Similarly, in the accounting area, many studies (ibid., pp. 7–8) have suggested, directly or indirectly, that little progress has been made toward the harmonization of accounting standards across nations. It is likely that different nations adopt different FERM tools and therefore subscribe to different financial products in dealing with perceived needs. This chapter looks at FERM in two countries, the United States and Mexico. These countries, which border each other, are currently moving toward free trade and yet have banking, economic, accounting and cultural roots that are radically different.

13.2 LITERATURE REVIEW

The development of an effective strategy for managing currency risk should proceed in three distinct stages (Cornell and Shapiro, 1988, p. 44). First, management must decide what is at risk; this requires an appropriate definition of foreign exchange risk. Second, it must clearly identify the objectives of its exchange risk management program. Third, having determined the extent of its exposure to currency risk and defined its objectives, management can then design a set of company-wide policies to achieve its objectives. This would include utilization of financial innovations to the extent to which these innovations are permitted and available.

How management copes with exchange risk can be related to the different types of risk. Discussions of the management of currency exposure are traditionally divided into three sections: (1) transaction, conversion, contractual or cash flow exposure; (2) translation, accounting or balance sheet exposure; and (3) economic, operating or long-term exposure. A detailed discussion of each of these exposures is contained in most major international finance textbooks, for example in Eiteman, Stonehill and Moffett (1992). A more theoretical presentation can be found in Jacque (1981).

Each type of exposure is likely to have its own optimum risk management practices involving an accompanying range of useful financial techniques and products. In turn, these risk specific practices may be conflicting and often a firm-wide optimum strategy must be sought within national and market constraints. This firm-wide optimum will carry its own unique demand for financially innovative products.

Financial innovation is defined in many ways. Keeley (1987, p. 118) defines it as a new financial product or practice, and also as the process by which new financial products and practices emerge. Desai and Low (1987, p. 114) build upon this distinction, defining innovation as a product that locates and fills a gap in the range of available products with such a gap definable in terms of existing product characteristics. More recently, Ross (1989, p. 544) describes financial innovation as a natural adjunct of the supply and demand of agency-constrained financial market participants, with the cost of marketing new financial products helping to shape the form of the new institutional features. Summarizing, one can define financial innovation in the following terms: given the ability to define a financial product in terms of its attributes or characteristics, an innovation is a newly engineered product whose definable characteristics are significantly different from those of any existing product and whose relative value can be readily determined or shown.

A plethora of factors has been suggested that affect either the demand for, or supply of (or the interaction between demand and supply forces) new financial contracts as a response to perceived FERM-based needs. A summary of the financial innovation literature is presented by Finnerty (1988, p. 16), who classifies these factors into the following eleven categories:

1. tax asymmetries that can be exploited to produce tax savings for one market participant or another;
2. transaction costs;
3. agency costs;
4. opportunities to reduce some form of risk or to reallocate risk from one market participant to another;
5. opportunities to increase an asset's liquidity;
6. regulatory or legislative environment or change therein;
7. level and volatility of interest rates;
8. level and volatility of prices;
9. academic work that resulted in advances in financial theories or better understanding of the risk-return characteristics of securities;
10. accounting benefits;
11. technological advances and other factors.

One possible oversight in this listing of factors is the level of competition in the financial markets. Most researchers in this area have cited such competition as a major factor in the increase in financial innovation over the past few years. The inclusion of competition as a factor completes the listing of important factors associated with financial innovation.

13.3 RESEARCH QUESTIONS AND METHODOLOGY

For the factors cited by the preponderance of the literature as primary environmental influences on financial innovation, that is, taxes, regulations, interest rate and price volatility, technological achievements and financial service industry competition, the United States and Mexico have been substantially different historically, and one would expect *a priori* different product profiles. Thus, the research question to be examined is: 'Do Mexican and U.S. companies differ in their FERM practices and if so how and why?' No specific research hypothesis will be tested; however it is anticipated that substantial differences in FERM, as measured by the utilization of certain FERM products, will be found.

A questionnaire survey of the 100 largest Mexican companies currently listed in Duns Latin American Guide was conducted. A total of fourteen responses was received. The questionnaire used was a derivative of an instrument developed for a similar study in the United States and already pre-tested and used in Jesswein (1992). The data collected for Mexico was compared with the results from Jesswein (1992), which included the responses of 173 U.S. firms.

Analysis was conducted on the behavior and perceptions of respondent companies on four parameters:

1. Use of FERM products and techniques.
2. FERM philosophy as it relates to: the need for active foreign exchange management; the need to cover particular risks; and perceptions of particular innovations.
3. The evaluation by the respondents of the relative importance of selected product characteristics in their decision to select particular FERM products. These factors were selected from the literature.
4. The perceived importance of particular factors in generating financial innovation in general. These factors were selected from the literature.

All questions were scored on a scale of one to five (the first question had a sixth category) with one indicating a high level of usage of a particular product or a high degree of agreement with a particular statement. The responses were then analyzed using t-tests and non-parametric ANOVA.

13.4 RESULTS

The mean and variance for each country on each question was computed and parametric and non-parametric t-tests conducted for each question. These were analyzed within the structure of the four questions described above.

Question 1

This question attempted to ascertain the level of utilization of different financial instruments used for hedging foreign exchange risks. The means of the responses to this question are summarized in Table 13.1. Substantial agreement was found on several issues. In both countries the variety of financial instruments and techniques used was low, with U.S. companies slightly more likely than Mexican companies to consider using newer financial techniques. In addition, companies in both countries showed higher utilization scores for more established and less sophisticated products such as forward and futures contracts.

Table 13.1   Consideration and utilization of selected FERM products (means of survey responses)

	U.S.	Mexico
Forward contract	1.5	3.1
Futures contract	4.0	2.7
Exchange-traded options	4.0	4.4
Futures options	4.4	4.6
OTC options	3.2	4.1
Synthetic option	4.3	5.2
Synthetic forward	4.3	5.4
Cylinder option	3.8	5.4
Participating forward	4.3	5.2

The U.S. companies in this sample tended to prefer the use of forward contracts, although over-the-counter (bank) options were also frequently used or considered for use. However, the balance of the listed products was not often even considered for use. Surprisingly, over-the-counter options had the highest variance, suggesting perhaps that they were a second step taken only by more sophisticated companies. For all products there was considerable dispersion in the level of utilization, indicating quite heavy usage among some companies and almost none among others.

The product most often used by Mexican companies was the futures contract, followed by the forward contract. The mean level of usage of both these products was significantly different between Mexican and

U.S. companies.* As with the U.S. companies other products were not often considered. Included in this latter group, the consideration of synthetic forwards, cylinder options and participating forwards by Mexican companies differed significantly from their U.S. counterparts. It is possible that this was due to lack of availability, a hypothesis confirmed by the emphasis placed by Mexican companies on availability as a factor in FERM product choice (see Question 3).

Question 2

This question dealt with the risk management philosophies and practices of U.S. and Mexican firms. The results of this section of the questionnaire are summarized in Table 13.2.

Table 13.2   FERM philosophies and practices (means of survey responses)

	U.S.	*Mexico*
Transaction exposure important	1.4	1.3
Innovation leads to sophistication	2.4	2.2
OTC options more flexible than exchange options	2.0	2.8
Translation exposure important	2.4	1.3
New products have little intrinsic value	3.2	3.6
Active risk management worth the effort	1.7	1.7
New products too complicated	3.6	2.8
Economic exposure important	1.8	1.3
Risk management more sophisticated since 1980s	2.0	1.7

On the question of type of exposure, both Mexican and U.S. companies tended to agree with the need for foreign-currency management to combat all three types of exposure. The area of highest concern to both Mexican and U.S. companies was the management of transaction exposure. For translation exposure, U.S. companies tended to be statistically significantly less concerned. Primarily, this may in part be traced to the U.S. ability to bury the results of foreign exchange translation losses in their retained earnings statements, whereas the Mexican companies, some of which have U.S. ownership, primarily use the temporal method and must report foreign exchange gains and losses on the income statement. Economic

---

* All significant differences are at the 0.05 level or lower.

exposure was also significantly more important to the Mexican companies than the U.S. ones. Essentially, Mexican companies have to deal with the fact that their own currency is declining on a regular basis, and so they appear to be willing to incorporate foreign-exchange strategies as part of their normal day-to-day business. The U.S. companies, having to some extent a wholly domestic market (inputs and outputs), can take a somewhat more aloof approach.

In general, Mexican companies strongly agreed that active foreign currency management was worthwhile. U.S. companies tended to take a more passive approach. Both groups strongly agreed that they had become more sophisticated in the management of foreign exchange.

Mexican companies generally found the new foreign exchange products to be too complex for general usage and this is seen in their choice of products. U.S. companies disagreed with the statement that new foreign exchange products are too complex, suggesting that they have adopted a more sophisticated approach to foreign exchange management. In addition there was considerable disagreement as to the appropriate vehicle. U.S. companies favored over-the-counter tailor-made options, while Mexican companies, placed their faith more in market-traded instruments and expressed the opinion that they did not feel that they could devise products more worthwhile than those being generated by the market.

In conclusion, the information discussed above suggests that Mexican companies have a stronger and broader commitment than U.S. companies to exchange management, but are less than comfortable with their own ability to design and use the more recent innovations.

Question 3

This question attempted to evaluate reasons for adopting various foreign exchange products based on a list generated from the literature. The mean responses to the issues raised here are found in Table 13.3.

The U.S. and Mexican companies had no significant differences in their rating of any one factor used in evaluating foreign exchange products. The Mexican companies rated total cost and compatibility with existing strategies as the most important factors but certainty of tax treatment and familiarity with the product were also considered factors of extreme importance (that is, a scalar score of less than 1.5). Availability of the instrument, familiarity with the underlying technology, compatibility with risk management philosophy and ability to

Table 13.3 Importance of factors used in evaluating FERM products (means of survey responses)

	U.S.	Mexico
Availability of instrument	2.0	1.6
Existence of upfront costs	2.5	1.8
Familiarity with product components	1.8	1.5
Risk of using new product	2.0	1.8
Familiarity with product technology	2.0	1.7
Certainty of tax treatment	1.9	1.5
Certainty of accounting treatment	1.9	1.8
Compatible with risk management philosophy	1.8	1.6
Compatible with accounting system	2.7	2.3
Compatible with hedging needs	1.6	1.4
Observability of prior usage	2.8	2.4
Identity of seller	2.3	2.1
Discussions in professional literature	3.1	2.7
Total cost to implement	1.8	1.4
Ability to cancel or close out	2.0	1.7
Existence of multiple uses of product	3.3	2.8
Flexibility of use of product	2.4	2.1
Overall 'value'	1.8	1.9

cancel or close a position were also considered very important (that is, a scalar score of less than 1.8).

The U.S. companies tended to place less importance on each individual criterion. The highest-rated factor for them was compatibility with corporate hedging needs. Familiarity with the product, compatibility with risk management philosophy, total cost and overall value were also considered very important (that is, a scalar score of less than 1.8).

Factors of little importance (that is, a scalar score of greater than 2.5) for both groups of companies were discussion of the product in the academic literature, and multiplicity of uses for the product. The U.S. group also seemed uninfluenced by the observation of previous successful applications and compatibility with existing accounting systems.

The results seem broadly in line with the literature, with the exception of studies that have suggested potential linkages of the impact of foreign exchange risk management decisions with the academic or practitioner literature, existing accounting systems, and multiplicity of use of the product.

Question 4

This section followed up on the individual manager's perception of the influence of factors on his or her decision with a general opinion question on factors likely to have affected financial innovation. The summary results of this section are found in Table 13.4.

Table 13.4　Factors associated with innovation in the financial markets (means of survey responses)

	U.S.	Mexico
Realization of tax advantages	2.1	1.5
Reduction of transaction costs	2.2	1.3
Reallocation of financial risks	1.9	1.7
Increased liquidity of funds	2.4	1.5
Avoidance of market regulations	2.8	2.3
Interest rate volatility	1.8	2.0
Exchange rate volatility	1.7	1.8
Advances in academic research	2.8	2.8
Realization of accounting benefits	2.7	2.3
Advances in technology	2.5	2.7
Reduction of agency costs	2.8	1.9
Financial industry competition	2.1	2.3

Costs (in this case opportunities to reduce costs) were seen as most important by the Mexicans and significantly less so by the U.S. group. The Mexican companies also strongly agreed that opportunities to increase liquidity and realizing tax advantages were important, significantly more so than their U.S. counterparts. Other relatively strong positions (that is, a scalar score of less than 1.8) included the opportunity to reallocate financial risk and the level and volatility of exchange rates.

The U.S. companies were less firm in their views but agreed most strongly with the need to reduce exchange rate volatility as a *prima facie* cause of innovation. The volatility of interest rates and opportunity to reallocate financial risk were also seen as significant drivers of innovation in financial markets (that is, a scalar score of less than 2.0). These secondary drivers were in concomitance with the Mexican perspective.

In terms of least important innovation drivers, companies from both countries discounted the effect of academic and financial research.

They differed significantly however, in that U.S. companies considered that agency theory and opportunities to exploit market regulation were ineffective in driving innovation, while Mexican companies did not. Surprisingly, the Mexican companies discounted that value of advances in computer and communications technology in the innovation process.

13.5 CONCLUSION

In moving south, U.S. companies will find a market that is as or more committed to FERM than they are. This market will be familiar, having many of the same products that the U.S. manager uses. However, the U.S. manager can possibly add a greater understanding and a range of products that may be useful to his Mexican counterparts. These Mexican managers, while sharing the U.S. managers' desire to avoid various forms of risk, will generally be found to be more cost-, regulation-, and taxation-conscious, factors that can only change as the economic and regulatory environment of Mexico itself changes.

Mexican managers expanding northward may have to reevaluate their repertoire of FERM techniques, but may be pleasantly surprised, given the relative inactivity of U.S. managers, at how well their previous experience has prepared them. It is the conclusion of this study that while the utilization and motivations for use of particular products may be different between Mexican and U.S. companies, the overall pattern of product usage and factors related thereto, is not significantly different between the United States and Mexico.

Part IV
Cross-Border and Industry-Specific Issues

14   Introduction

The chapters presented in this part examine the role and significance of the North American Free Trade Agreement in relation to cross-border and industry-specific issues. In Chapter 15, James Lane studies the issue of labor turnover and job training in the maquiladora industry. Using exit polls of maquiladora workers, he measures the impact of the training that workers receive in these plants on employee turnover in each plant. Lane's survey of a sample of 121 maquiladora plants in the San Diego area resulted in 39 usable responses. His research lead him to conclude that the correlation between labor turnover and training is function of the type of training. For example, he found no correlation between labor turnover and classroom training. On the other hand, he found negative correlation between labor turnover and on-the-job training and positive correlation between employee turnover and other training.

In Chapter 16, Jane LeMaster and Bahman Ebrahimi examine Ronald Ayers' model of human capital investment and study its applicability to the U.S.–Mexico border region. They use an internal rate of return method of analysis to test Ayers' argument that there may be less incentive for human capital – defined as the investment in an individual's training or education made by either the individual or his employer – in border regions because of the political risk associated with borderland economics. LeMaster and Ebrahimi develop four propositions and test them against Ayers' model and conclude that while political risks associated with the economy of the U.S.–Mexico region are logical and appropriate, such risks should not be used as the sole or even major determinant of human capital investment. Among the other factors that LeMaster and Ebrahimi propose are time-orientation ('Persons oriented toward immediate gratification will be less likely to invest in human capital') and age of the employee (because the yearly marginal earnings are the same for younger people as for older people, the former are more likely to invest in human capital). The authors also argue that the costs and availability of human capital and earning differentials should be included in any analysis of investment in human capital.

In Chapter 17, Lawrence W. Nowicki studies the Mexican export processing zones (EPZs, more commonly known as maquiladoras) in

Sonora, Mexico, with the goal of determining whether they can be used as the future outlines of the functioning of the North American Free Trade Agreement. The chapter contains the findings of the author's questionnaire and interview survey of eleven separate maquila plants in five different cities in Sonora. Using the findings of his survey, Nowicki compares border maquilas with those located in the interior of Mexico using variables ranging from work-force characteristics to workers' attitudes toward the concept of maquiladora; from investment location factors to productivity and turnover; and from the workers' education level to migration trends.

In Chapter 18, Nancy Wainwright addresses the question of trade relations between the United States and Mexico, particularly insofar as the fruit and vegetable industry is concerned. She traces the recent events, particularly congressional hearings, that have dealt with the fruit and vegetable industry and concludes that 'a critical issue is how to prevent a market collapse in a particular agricultural product, an issue that the United States has been addressing since the Great Depression and continues to address in the free trade and GATT roundtable discussions'. Wainwright also reminds her readers that 'improvement in trade relations would benefit all three countries whether or not a free trade agreement is negotiated and whether or not GATT is able to resolve the issue of subsidized agriculture'.

The apparel industry is the subject of Chapters 19 and 20. In Chapter 19, Sandra Forsythe, Mary E. Barry and Carol Warfield examine the broader issue of the global competitiveness of the North American apparel industry; while in Chapter 20 Kathleen Rees, Jan M. Hathcote and Carl L. Dyer study the more specific question of the effects of NAFTA on the apparel industry in the United States. Forsythe, Barry and Warfield maintain that as a result of the relaxation of trade restrictions, economic reform in Mexico and NAFTA, increased opportunities will exist for greater cooperation among North American manufacturers in the apparel industry. They argue that 'complementing the present Canada–U.S. FTA with the unique advantages offered by Mexico, a globally competitive apparel manufacturing industry is possible. Indeed, the combination of labor and market power could result in an apparel industry that will compete effectively in a global market'. To that end, Forsythe, Barry and Warfield devote the rest of this chapter to address market opportunities for increased apparel trade between the United States and Mexico and the potential for a globally competitive apparel industry in North America.

Rees, Hathcote and Dyer, on the other hand, compare the textile and apparel industry in the United States and Mexico, discuss the specific sectors within the United States which will benefit, or suffer, from NAFTA, and provide a differing view of the 'global impacts and ramifications of the agreement'.

15 Maquiladora Employee Turnover and Job Training

James M. Lane

15.1 INTRODUCTION

Exit polls of maquiladora workers have indicated the value the worker has placed upon the training they have received at the plants. Yet there have been no studies examining the effect of that training upon worker turnover. This chapter will examine that relationship in an effort to determine the possibility of using training programs as a method of slowing turnover.

The 'maquiladora' is a manufacturing/assembly industry created over 25 years ago by Mexico to alleviate some crucial social and economic problems. That the maquiladora (maquila) is successful in economic terms for Mexico has been well-documented. Many statistical studies have been made of maquilas at the macro levels of economics and trade. Other research areas include those in the social sciences, especially concerning the emergence of the Mexican woman as the chief source of labor in the maquiladora work force; her arrival, in many cases, as the head of the household in economic terms; and the socio-cultural turmoil which has occurred because of these and related events.

In these studies, however, there is a lack of academic or research materials oriented toward making a business decision concerning production alternatives and examining the maquiladora as an alternative in that business decision.

In some studies, the findings are contradictory – possibly refecting the viewpoint of the study's funding source; or because of differences in the study's actual orientation.

Of the small number of pragmatic studies useful to an executive when evaluating the maquiladora, most have concentrated on economics, employment, the job-loss/job-gain controversy, turnover problems and environmental issues. Others have looked at site location decisions within Mexico as an alternative for total off-shore production, transfer of technology, legal issues, production sharing and, very recently, return-on-investment changes as the result of establishing a maquila.

There have been many macro-level studies that have included the maquiladora as one element. There have been studies, theoretical and case, on production sharing, which may be pertinent; in a few instances historical studies have determined why a company picked one location over another; and case studies have been conducted on setting up a plant utilizing high technology in an interior region of Mexico.

Studies on certain key elements such as turnover, migration and labor have been done, but usually these have been approached and discussed from the social sciences viewpoint. Although that orientation has some pertinence to the executive, the findings are not couched in the pragmatic terms needed by the executive in a business context.

This chapter expects to illuminate further the turnover problem as it is examined across a series of management variables and in the context of the executive's opinion of turnover. Further, it will allow that executive to evaluate the turnover issue as experienced by his predecessor and colleagues. It will furnish the executive with some additional guidance.

15.2 REVIEW OF THE LITERATURE

Turnover rates since the late 1970s have averaged between 10 per cent and 35 per cent per month, which is much higher than in competing Pacific Rim countries. Over 75 per cent of all turnover occurs in the first three months of employment. This has the effect of lowering productivity and employee morale, increasing downtime and incurring extra costs associated with additional recruitment and training efforts (U.S. International Trade Commission, 1988, pp. 8–11).

Business International Corp. (1986, pp. 397–8), counseled firms that the turnover rate ranges from 30 to 130 per cent per year in different regions for maquiladoras. They suggested expecting turnover rates of 20 per cent per month, especially toward August, when some student workers return to school. Business International advised extensive screening to help identify problem employees (also Lucker and Alvarez, 1985), having the right location, using trade unions to help stabilize the work force, offering competitive benefits and worker perks and using maquiladora associations to cooperate in an effort to reduce turnover.

A Wharton study warned that even though high turnover rates could be avoided by locating new firms in the interior of the country, this move would imply a loss of closeness to the U.S. market and the need to use an infrastructure that may be inappropriate in some localities (1988, p. 2).

The labor shortage is expected to worsen soon because of continuing expansion of the United States companies along the border. For the first time, United States companies are recruiting workers further south and beginning to locate the maquilas in cities in the interior of Mexico (Gilbreath, 1986, p. 10).

And from another study:

Although the maquiladoras have provided a significant source of new employment for the 'fronterizos', some critics point to the negative results of this form of industrialization. Male unemployment has been only slightly reduced, primarily due to the practice of staffing the factories largely with women (Martinez, 1983, p. 14).

Lucker and Alvarez carried out a study of the links between certain aspects of workers' backgrounds, their personalities and turnover. Results suggested that working longevity could and should be proactively managed at the time of hiring (1985, p. 8).

There are problems in the industry, most of which stem from the rapid expansion of the industry itself. Included in these are high employee turnover rates and absenteeism (Clement and Jenner, 1988, p. 3).

Workers with a seasonal-work background are high turnover risks. Locating closer to residential areas reduces transportation subsidies as well as turnover. Perks have to meet or exceed those of competitors or workers will move, and extra perks should be used to reward top performers.

Stoddard (1987, pp. 45–60) responded to charges of high labor turnover being an exploitative corporate strategy by stating that corporate managers are extremely unhappy about turnover, being an added cost in their production budget (1987, p. 47). Turnover is seen as a costly problem rather than a fortuitous strategy to control workers.

According to Gambrill (1986) wage raises after 1982–3 slowed labor turnover in the maquilas, indicating wages may be used as a control.

15.3 RESEARCH DESIGN

The respondents to this survey were either the executives involved in the actual decision-making process which resulted in their operating a maquiladora in Mexico, or the most knowledgeable person of the process available. In every case, these were executives who were able to provide historical overviews and analyses of the operation – a perspective not available through other sources.

The San Diego-based sample was used to facilitate the interview process used to collect most of the data. The group of maquiladoras having parent firms in San Diego represented 157 of the 413 maquiladora firms licensed in June 1988 by Secretaria de Comercio y Fomento Industrial (SECOFI). Each parent firm was limited to a single response no matter how many maquilas they operated. This reduced the study's sample to 121 firms, which produced a final response of 39 firms with usable data.

The survey was carried out either by a face-to-face interview or by a questionnaire mailed to executives meeting the criteria.

The survey covered the following information about each maquila: (1) Mexican Industrial Classification Code, (2) ownership mode, (3) plant output in terms of products and units, (4) information on Mexican employees' training programs and turnover and (5) open-ended responses on a variety of subjects.

15.3.1 Reliability, Validity and Testing

Once the concept of the study had been defined, the literature was searched for possible surveys to use. Nothing was found which addressed the concerns of this study. A researcher-constructed survey was then developed.

The survey was submitted to a group of 15 executives associated with the maquiladora industry, and their opinions were sought on what the survey would measure in reality, whether it would collect the necessary data, and what changes or modification would be advisable. Following the Delphi-type approach, the final survey was again checked using a sample of 15 in Sonora, Mexico, and it was then confirmed.

Assumptions

The following assumptions were made: the data gathered was honest and accurate; the respondents were the population sought; and the source documents used to determine the sample population were reasonably accurate.

Limitations

The following limitations apply: most firms did not have, or would not provide, data related to the cost of labor turnover, training budgets, or plant output specified in monetary terms of any kind. Most firms did not have a highly structured training program, so hours spent in training could not be accurately determined.

15.4 FINDINGS

Labor turnover was examined according to the employee-training methods utilized at the maquiladora operation. The categories were: on-the-job training; classroom training and other training. Respondents picking 'other' were asked to specify the training used. None did so.

There was no correlation between labor turnover and classroom training, indicating the classroom has little to do with turnover factors.

The correlation between the labor turnover rate and other training methods was positive, indicating that greater turnover was associated with methods of training other than on-the-job or classroom training:

$$\text{(Turnover/other training)} \quad \frac{r}{.4508} \quad \frac{N}{(32)} \quad p = .008$$

On-the-job training showed a negative correlation with labor turnover, indicating that increased on-the-job training was associated with lower turnover. However, no determination was made of an optimum amount of training in either arena:

$$\text{(Turnover/On-the-job training)} \quad \frac{r}{-.4531} \quad \frac{N}{(33)} \quad p = .009$$

The final section of the survey was a series of open-ended questions concerning the positive or negative aspects of the maquila, and trends the managers expected to see both in their particular industry and in the industry as a whole with regard to labor and turnover.

The negative labor aspects of maquila operation were given as turnover rate (16), infrastructure (10), lack of highly qualified technicians and managers (6) and wage increases and cultural conflicts in the workplace (5). Additional negatives were noted for a lack of skilled labor (3).

U.S. managers indicated the following labor trends were expected in Mexico: quality labor availability will decrease (14), rampant or fast growth of maquilas (9), labor costs will rise (8), turnover will increase (7), turnover will decrease and better-educated workers will be available (4).

The data is mixed on several counts, with some predicting more workers and others less; some predicting more turnover and others less.

This data, although widely scattered with opposites showing comparable frequencies, may not be statistically validated; but it does give the opinions of the U.S. managers and should serve as both a positive reflection of what the maquiladoras should expect while posting a warning sign about what lies ahead.

15.5 CONCLUSIONS AND RECOMMENDATIONS

An opening note to this chapter acknowledges the rapid expansion and change in the area of maquiladoras which has very likely dated and possibly made obsolete some of the findings of the study presented in this chapter.

The purpose of the study was to examine personnel turnover within the maquilas, to find any determinants of turnover, and to examine turnover in terms of plant size.

Another area of concern was to examine the current managerial views on the positive, negative and future trends of the industry with regard to labor training and turnover.

To accomplish this research a survey was developed and tested, then mailed. Interviews were undertaken to complete the gathering of the data. Of the original 121 firms targeted, 34 no longer exist. Finally, a total of 39 usable responses was obtained.

The reason for the low return may have been due to the events taking place at that time – the Mexican government was conducting its own survey on the hazardous wastes being generated by the maquiladoras and the handling of those wastes. This environmental study was an official government undertaking and had the possibility of affecting the future of the individual maquilas, making response to that survey a priority.

In an attempt to understand better some key variables as predictors of labor turnover, variables indicative of the plant's size were tested. They were output, plant area and total Mexican labor. A common view is that as the plants get larger, turnover increases. This view was not supported by the study. No correlations were found, indicating these variables were not indicative nor predictive of labor turnover within the study population.

Turnover is the most critical problem in the maquiladora industry and should be a key thrust of researchers in an effort to help bring turnover rates down. Several methods of mitigating turnover are being attempted. These efforts include providing housing, transportation, cafeterias, medical clinics, nurseries, bonuses for attendance and production, and decuation refunds, as well as establishing the plant away from the border. The only significant method documented so far has been to move into the interior of Mexico, but this is not feasible for all operations considering the maquiladora option. Therefore methods need to be developed to deal with the turnover problem, especially in the border areas.

Two variables were shown to be significantly correlated with labor turnover in Mexico. First, on-the-job training had a negative correlation, indicating that a higher amount of time spent in on-the-job training has the effect of reducing turnover.

The study did not gather any data which would help to determine whether there is an optimum level of on-the-job training to reduce turnover, it only determined that the relationship exists. Classroom training was not significant, which may be due to the few numbers of firms using this mode of training.

The on-the-job training finding supports other research which has indicated that the typical worker in a maquila views training as being the greatest asset of the experience.

The second variable was the significance between 'other' training methods and turnover. Here the correlation was positive, indicating these unspecified training methods cause an increase in the turnover at those maquiladoras.

Further research is suggested in the area of labor turnover. Several areas which need examination include determination of the optimum training hours for a given industry in order to balance increased training and costs versus lower turnover savings; the performance of the maquiladoras based upon their individual industries, the mode of ownership and location; and definition of what was meant by 'other' training methods.

The data used in the study was from a small sample and the conclusions should be tested in other border and interior regions of Mexico.

16 Analysis of Ayers' Model of Human Capital Investment with Political Risk and its Application to the U.S.–Mexico Border Area

Jane LeMaster and Bahman Ebrahimi

16.1 INTRODUCTION

In this chapter Ayers' model of human capital investment will be discussed and its applicability to the Texas–Mexico borderland will be investigated. Ayers (1988) suggests that because of the political risks associated with borderland economies there may be less incentive for human capital investment. Human capital investment is defined as the investment in an individual's training or education made by either the employer or the individual. Following the analysis by Flanagan *et al.* (1989), an internal rate of return method of analysis will be applied to Ayers' model. Four propositions will be presented and discussed.

The Texas–Mexico border area has high potential for industrial development because of its high unemployment rate, low per capita income and low levels of education. The low levels of education exist in spite of the four major universities located along the Texas border within at least 15 miles of the Mexican border. February 1991 unemployment figures show a 14.1 per cent average unemployment rate for the Rio Grande Valley of Texas from El Paso to Brownsville. The highest unemployment rate, 19.6 per cent, is in the McAllen/ Edinburg/Mission area (*Dallas Times Herald*, 14 April 1991). The Texas–Mexico border area has been slow to develop industrially, despite its high potential for development, and it remains primarily a labor-intensive agrarian economy.

An inherent problem in the borderland for industrial developers is the associated political risks caused by the symbiotic relationship between the two countries (Ayers, 1988). The area's economy is affected by the import/export laws of both countries, their tax structures, and the myriad of detailed regulations on the flow of goods and services across borders. The border regions are affected not only by the economies of both countries, but also by all the regulations between the two countries, as well as immigration – both legal and illegal. Ayers (1988) suggests that because of political risks associated with borderland economies, there may be less incentive for human capital investment. This is due to the fact that the certainty of return on that investment is not high enough. The purpose of this chapter is to provide a critical analysis of Ayers' Human Capital Investment Model, which includes an adjustment for political risk, and to determine the applicability of the model for the borderland region between the Lower Rio Grande Valley in Texas and the northern border of the state of Tamaulipas in Mexico.

16.2 DEMOGRAPHIC DESCRIPTIONS OF THE BORDERLAND REGIONS

Two contiguous counties in South Texas – Hidalgo and Cameron – are the focus area for application of the model analyzed in this paper. There are two major highways leading to the Lower Rio Grande Valley in Texas – Highway 281 into Hidalgo County and Highway 77 into Cameron County. Other than a small port at Brownsville, distribution into and out of the area is by truck. The bulk of income is derived from citrus fruit, cotton and vegetables. Recent years have seen somewhat of a 'boom' in the tourist industry, with people from the north and mid-west going to the valley for the winter (Winter Texans). Population growth rates are among the highest in the nation (Hansen, 1981). Over half of the population lives below the poverty level and less than half graduate from high school.

On the Mexican side of the border, the two major cities of interest for this chapter are Matamoros, which lies across the Rio Grande River from Brownsville, and Reynosa, which lies across the river from McAllen. In recent years, presumably because of increasing economic difficulties in Mexico and increasing immigration into the United

States, growth in the border towns has been more rapid than in the interior. It is clear that the borderland of South Texas and Mexico is an area with low incomes, low educational levels and low industrialization.

16.3 AYERS' MODEL

Ayers attempts to address the issue of human capital investment from the demand side of the economic equation. He states that there is 'a naive assumption . . . that providing greater access to education will necessarily result in significant increases in the desire to take advantage of the improved access' (Ayers, 1988).

Human capital investment is generally defined as the investment in an individual's training or education made by either the employer or the individual. There are three different types of investment that workers make in themselves: (1) education and training, (2) migration and (3) search for a new job. Presumably any investment is made with the expectation of increasing future earnings. Economists conceptualize human capital investment as a term that embodies 'a set of skills that can be "rented" out to employers' (Flanagan *et al.*, 1989). Ayers defines the human capital model as representing the 'worker as an investor in self-education, who follows the rule of rational investment, which says that one will invest in an increment of human capital only if the expected return from the increment exceeds the expected cost of producing it' (Ayers, 1988).

Ayers defines political risk as the unique problems associated with border areas and gives examples of economic events in Mexico that would be considered political risks. These risks include devaluations, exchange controls, import and export restrictions, changes in taxation and changes in labor policy. These types of economic events in Mexico also affect the profitability of industries in the United States. Although organizations may be able to build-in buffers to accommodate such events, the individual workers may not have the kinds of options that would afford them the choices to accommodate political risks. 'The root of the problem', according to Ayers, 'is that human capital cannot be separated from its owner'.

When discussing political risk from the perspective of the organization, strategies for dealing with political risk are complements. Individual workers' strategies for dealing with political risk, however, are substitutes. Ayers contends that strategies for individual workers living in borderland areas consist of (1) not making the investment in human capital at all, (2) limiting the amount of investment, or (3) simply moving away from the border area and seeking employment elsewhere.

There are a number of factors that need to be considered when long-term returns on an investment in human capital are calculated. These are the expected future earnings and the time period over which they will occur, along with the expected costs of that investment. This leads to a simple present-value calculation of expected future earnings and anticipated costs. Ayers' model incorporates this simple present value calculation with a factor for political risk to achieve the net capital value (NCV) of a human capital investment. Ayers' formula for calculating NCV is:

$$\text{Max NCV} = \sum_{t=0}^{n} \frac{\sum_{j=1}^{m} (P_j E_j)_t}{(1 + i + u)^t} - \sum_{t=0}^{n} \frac{\sum_{k=1}^{s} (P_k C_k)_t}{(1 + i + u')^t}$$

where P_j = probability of earnings stream E, E_j = earnings stream, i = personal discount rate, u = earnings political risk premium, P_k = probability of cost stream C, C_k = cost stream, u' = cost political risk premium, m = total A of earnings outcomes and s = total A of cost outcomes.

The left-hand side of the equation is the maximum net capital value (NCV) of an investment in human capital. The first summation term on the right-hand side of the equation represents the present value of the earning stream over the collective employment lifetime. Included in the denominator is the factor u, which represents the earnings' political risk premium. It is the factor included for the political risk associated with borderland employment. The u factor is included in the denominator because there is an inverse relationship between political risk and the predictable value of future earnings. The political risk factor u takes on positive values in border cases, because of the greater potential for higher political risk than that of an otherwise *ceteris paribus* worker not on the border.

The second summation term on the right-hand side of the equation represents the present value of human capital investment costs over collective employment lifetime. This term is subtracted from the first term. The numerator of the second term represents the probable cost factors associated with human capital investment. The denominator represents the discount rate and u' the cost of the political-risk premium. Ayers contends that u' will have a negative value because of the inflated cost of human capital along the border areas relative to that which would be expected with no political risk.

16.4 EVALUATION OF AYERS' MODEL

The general structure of Ayers' model is logical in that the maximum net capital value is the difference between the present value of expected future earnings less the present value of the costs of future human capital investments. Also, the inclusion of the political-risk premium in the expected earnings curve is logical because certainly future employment and earnings will be determined by all of the swings in the borderland economies, as noted earlier. Figure 16.1 shows Ayers' argument that the demand curve for human capital investment of the borderland worker will be to the left of the demand curve for human capital investment of the non-borderland worker (Ayers, 1988).

Costs (represented by the second term on the right of the equation) of attending a trade school, technical school, junior college or university are not determined by the border economy; i.e. tuition, books, and transportation are determined by other factors. The variables of rent and food that might change as a result of border economic variations would have to be paid whether one was investing in human capital or not. Rent and food are part of the standard costs of living, not a factor in human capital investment. The factors that would influence these returns would be the psychic costs involved with the inconvenience and uncertainty of deferring present income to obtain education or training.

Figure 16.1   Diagram of Ayers' argument

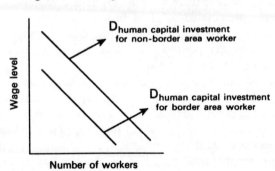

Ayers' concluding remark in his analysis is that 'other things being equal (such as access to education), it is rational to invest less in oneself if one resides along the border'. It is apparent that excluding the *u'* factor from the second term would not change Ayers' conclusion because of the presence of the *u* premium in the first term, which decreases the net capital value. Ayers suggests government intervention as a way of reducing political risk and raising the return on human capital investment. The second term, costs of political risk, seems to have been neglected. In addition to reducing political risks, public policy could also include reducing the costs of human capital investment through scholarships, low-interest loans, special low-tuition for border zones and similar strategies. Much of the literature concerning human capital (Hu, 1976; Manning, 1975, 1976, 1982; Razin, 1972) discusses optimal government educational policies from the perspective of the '. . . production function for human capital, the distribution of learning abilities in the population, the skill intensity of internationally traded commodities, and so on' (Blomqvist, 1986).

Following the analysis of Flanagan *et al.* (1989), one can adopt the internal rate of return method of analysis. To do this Ayers' model must be modified by setting the two terms on the right-hand side as equal and calculating the E_j terms as the marginal earnings that occur as a result of capital investment. Then one can ask the question: 'What rate of return would make the training or education investment break even?' Clearly it becomes apparent that any decrease in the political risk factor will increase the rate of return. This approach has the benefit that an individual can investigate his capital-investment decisions more objectively and quantitatively. That is, one person may be satisfied with a modest 5 per cent return on his or her human capital investment and another person may demand a 20 per cent return. If the analysis of internal rate of return turned out to have a breakeven point of 10 per cent, clearly the first person would accept it and the second would reject it.

Ayers' political risk factor could become time-dependent if one had a rationale for saying that the political risk factor will increase or decrease in the future. There may be strong political indications that the border area may become less restrictive on commerce between the two countries as a result of the free trade agreement now being discussed between the United States and Mexico (Vernon and Wells Jr., 1991, Polyconomics Inc., 1990). This possible time-dependence of the political-risk premium could be well accommodated by Ayers' model with the modifications noted above. In order to indicate change in time dependency, the equation could be differentiated with respect to political risk and time. If human capital investment changes with time, the slope would be equal to the differentiation. If the investment did not change with time the slope would be zero.

Applying the Flanagan *et al.* (1989) analysis of human capital investment to the Ayers model, we arrive at the following propositions:

1. Present-oriented people are less likely to make a human capital investment.
2. Primarily, younger people will make more human capital investments.
3. Human capital investments will increase as the costs decrease and availability of those investments increase.
4. Human capital investments will increase as industrial development in the border region increases the gap between earnings of higher trained and unskilled workers.

16.5 DISCUSSION OF THE ABOVE PROPOSITIONS

16.5.1 Present Orientation

Persons oriented toward immediate gratification will be less likely to invest in human capital because of their lack of interest, or inability to be interested, in future potential earnings. Low per capita income along the border regions also forces people to be more present-oriented than they might be if they had abundant economic resources. A good example is families on welfare who are not likely to make investments in human capital such as education or training. Adding political risk factors further encourages present-orientedness. Another factor contributing to present-oriented thinking is psychic stress. Psychic stress is a subjective term used to describe the stress one feels when considering moving to a new location, giving up friendships and relationships, or investing present assets for future benefits, particularly if those assets are limited.

16.5.2 Age

Given that the *yearly* marginal earnings are the same for younger people as for older people, the present-value calculation gives a greater rate of return to a younger person. For instance, younger people are generally the ones who go to college. Although there is ordinarily an earnings differential for experience, usually held by older persons, the upper limits of earnings are higher for the person with more education and training. The experience factor discourages the older worker from going back to school because there are higher opportunity costs associated with the potential loss of their higher earnings. If the political-risk factor is present for a younger person it decreases the present value of the discounted future earnings. So, the people who are more likely to undertake training, the younger people, have a smaller incentive to do so because of the political-risk factors associated with living in border regions.

Shaw (1989) derived a model of joint determination of labor supply and human capital investment and found labor supply to be differentially affected as individuals become older. The contention is that policy changes, such as changes in the tax structure (a political-risk factor), would have distributional consequences because the younger worker is less responsive than the older worker.

16.5.3 Costs and Availability

The relationship between cost and investment in self is indirect – the lower the cost, the greater the rate of return. The availability factor relates more to stress – the psychic cost – than to actual capital investment. The psychic costs are probably greater for older workers and workers with less skills or abilities and hence act as a deterrent to investment. Therefore both lower costs and greater availability would increase the likelihood of an individual deciding to make a human capital investment by participating in training or education. As mentioned above when discussing the age factors, the costs of training or education are greater for older workers. Also, as noted earlier, costs and availability of human capital investment strategies with respect to food and rent are factors unassociated with political risk, as they would be incurred whether or not one invested in human capital.

16.5.4 Earnings Differentials

Industrial development increases the demand for workers with specialized skills. The larger the organization, the more specialized the skills required. A neighborhood welding shop might require a person who can weld dozens of kinds of materials, whereas a company that manufactures stainless steel vacuum valves requires workers who are exceptionally good at high-quality welding of one specialized material. In an agricultural economy with only a small number of industrial organizations, there is little need for a lot of specialized training or a large, highly educated workforce. As economic development occurs, the demand for a trained workforce increases, and hence the marginal earnings increase between untrained and trained people. This demand increases the incentive for human capital investment. A symbiotic relationship then develops between educational and training institutions and industrial developers – they need each other. Increase in industrial development enhances the rate of return on the human capital investment that an individual might make in an educational institution. It is not unusual for a company to send a large group of workers for specialized technical training, or even to pay their tuition and fees for attending classes at a local junior college or university; that is, the company (firm) makes the human capital investment. The gap between trained and untrained workers in a situation where numerous plants have shut down due to changes in the economy – devaluation of the peso, import/export regulations – would cause a severe overabundance of trained workers and hence a small differential in salaries of trained versus untrained personnel. The difference, however, may be in being employed or not being employed. Therefore the perceived political-risk factor would be high and this would discourage investment in human capital by the untrained worker or additional investment by the trained worker.

16.6 CONCLUSIONS

Ayers' model, which allows for political risks associated with the borderland economies of South Texas and Mexico, is logical. The model does not allow for other factors associated with border areas; that is, the melding or perhaps creation of a new culture unique to symbiotic relationships between a developed country and a developing country. Without testing the model with actual data it is difficult to determine the potential magnitude of the numbers. Further, it is not clear that the political-risk factors described by Ayers can actually explain the greatest amount of difference. The model does not fully account for the vague intuitive psychic risk for potential economic boom or bust. Snow and Warren (1990) highlight the importance of acquiring empirical evidence regarding the effect of risk on human capital investment.

Further, it is not clear how Ayers arrived at the risk potential of the borderland worker being analogous to the risk factors of multinational firms. If reliable forecasts could be made regarding political-risk factors, and the model differentiated to reflect the dynamics of the political-risk factors, then the model would become very useful for determining whether or not to invest in human capital. The model is weak; however, given the terms of the model as Ayers has presented them along with the assumptions of borderland economies, the model can be applied to the region between the Lower Rio Grande Valley in Texas and the northern border of the state of Tamaulipas in Mexico. Empirically testing the model on actual data will be the best test.

17 Mexican EPZs as an Indicator of the Future Outlines of a NAFTA: The Case of Sonora

Lawrence W. Nowicki

'It was the best of times, it was the worst of times . . .'
Charles Dickens, *A Tale of Two Cities*

17.1 INTRODUCTION

Export processing zones (EPZs) have a dual nature: they are both an administrative instrument for providing free-trade status to a nation's manufactured exports as well as industrial parks specialized in manufacturing for export (World Bank, 1992). In the absence of an economy-wide, 'market-friendly' (World Bank, 1991) environment for growth – a favorable entrepreneurial climate, macroeconomic stability, extensive external links and broad state support for human capital and infrastructure – EPZs have been employed with varying success by less developed countries (LDCs), including several of the most protectionist, inward-oriented ones, as a selective policy tool in their attempts to earn foreign exchange, increase employment and ultimately induce industrialization. The use of EPZs to shift to a more open, competitive, free-trade strategy for growth and domestic development has been characterized as a 'post-Listian' breakthrough to the extent that it represents a radical departure from the restrictive strategy of import-substitution industrialization (ILO/UNCTC, 1988, p. 155). Perhaps the term 'anti-Listian' might be more appropriate to describe the 180-degree policy shift that the EPZ-assisted open border strategy of industrialization can represent.

The 'anti-Listian' free trade arrangements of Mexico's EPZs date back to the Border Industrialization Program of 1965. At that time, U.S.-made parts, machinery, equipment and even raw materials were first admitted into an extremely protectionist Mexico on a duty-free, in-bond basis on condition that, once assembled or produced, the resulting finished or semi-finished goods be exported out of Mexico, thus earning foreign exchange and creating jobs, especially in the chronically depressed U.S.–Mexico border region.

Mexico very exceptionally allowed 100 per cent foreign ownership of such inward investment. U.S Tariff Items 9802.00.60 and 9802.00.80 (806.3 and 807 prior to international harmonization in 1989) allowed duty-free reentry into the U.S. of goods assembled in Mexico from components originally made in the United States. Only the value added outside the United States – essentially by Mexican labor and non-U.S. parts – was subject to U.S. duty.

Mexican EPZs are known as maquiladoras or industrial parks. Individual in-bond plants, whether located inside or outside a given maquiladora, are termed maquilas. Successive Mexican governments have drafted legislation and decrees since the 1960s encouraging the spread of maquilas to the interior, and for good reason: these plants created most of the country's new jobs in the 1980s, directly and indirectly employing more than a million workers, including some of Mexico's most highly skilled labor. They have accounted for 80 per cent of the economy's manufactured exports, prevented default on the country's massive foreign debt, and are now the largest provider of foreign-exchange earnings after petroleum (Drucker, 1990; Fatemi, 1990). For observers such as Drucker, these results are the fundamental explanation of Mexico's policy reversal and decision to negotiate a free trade agreement with the United States.

A North American Free Trade Agreement (NAFTA) with Mexico can be interpreted as a 'wholesale extension of the maquiladora concept' (Hufbauer and Schott, 1992, p. 102). Macroeconomic forecasters, such as Wharton Econometrics (Vargas, 1991), have predicted that the maquiladora industry will serve as a blueprint for NAFTA becase of the abovementioned free trade arrangements that the industry's participants – mostly U.S.-based but also Asian- and European-based firms investing through their U.S. subsidiaries – already exploit in order to be cost effective and survive.

Why have the majority of the 100 largest U.S. firms invested in Mexican EPZs, including General Motors with 30 maquilas and General Electric with 13 (deForest, 1991; Michel, 1988), and what impact has this investment had on directly employed Mexican labor?

The first wave of maquilas essentially involved U.S. parent firms 'sharing' their labor-intensive assembly operations with their 100 per cent-owned Mexican subsidiaries. The dominant postwar Taylorist–Fordist model of scientific management and mass production, involving the separation of the tasks of conception and engineering by highly skilled labor from the tasks of execution by unskilled labor, enabled cross-border production-sharing to develop (Nowicki, 1982). As competition from lower-cost foreign producers intensified, the maquila concept permitted significant short-term cost saving, especially following the plunges in value taken by the Mexican peso after 1982. But today's maquilas are increasingly capital-intensive, applying flexible, post-Fordist lean production methods (Wilson, 1990). Maquila pioneer RCA at Ciudad Juarez, for example, now has robots, as maquilas are transformed from mere cost centers into the first examples of what integrated North American production will look like. These supply-side issues are analyzed below.

Although Mexican workers do have a social safety net, the mass consumption of consumer durables and the upgrading of basic infrastructure that also accompanied postwar mass production in the industrial countries have been lagging for maquila workers, creating the imbalances studied below, and which have been described as typical of the 'peripheral Fordism' of less developed countries (Lipietz, 1986).

Charles Dickens' famous opening lines in *A Tale of Two Cities* could well summarize the existing literature on foreign investment and EPZs in the border state of Sonora. The five-year MIT study of the world automobile industry found that Ford's plant in the Sonoran interior at Hermosillo 'had the best assembly-plant quality in the entire volume plant sample, better than that of the best Japanese plants and the best North American transplants' (Womack, Jones and Roos, 1990, p. 87). This plant has been termed a 'harbinger of future efficiency' in the event of an integrated, barrier-free North American market (Hufbauer and Scott, 1992, p. 220). Its workers, who were found to be 'young and eager to learn' (Business Week, 16 March 1992), 'embraced lean production with the same speed as American workers at the Japanese transplants in North America and at Ford's own U.S. and Canadian plants' (Womack, Jones and Roos, 1990, p. 265).

At the Sonoran border, the World Bank has cited the privately-owned EPZ of R. Campbell at Nogales as an example of the 'influential' role in world production-sharing played by Mexico's EPZs (World Bank, 1992, p. 26). This Nogales EPZ is recognized as the world innovator in shelter operations, whereby a zone contracts newcomer firms to manage their operations.

At the same time, the 'abysmal living conditions and environmental degradation' found at Nogales (Nazario, 1989), were said to 'rival any of the well-publicized disasters of the worst Stalinist regimes' (Kirkland, 1991). Nogales has been singled out as a prime example of the 'hope and heartbreak' (Tolan, 1990), 'boom and despair' (Nazario, 1989) that have coexisted along the U.S.–Mexico border. The data presented is based upon the assumption that NAFTA would eliminate the Mexican tariff on U.S.-made parts incorporated in maquiladora products that would be sold in Mexico rather than exported. Firms seeking to serve the Mexican market as well as the U.S. and other foreign markets through their maquila operations would thus have an incentive to shift production to interior plant locations in order to be closer to their Mexican clientele and to improve access to the country's pools of labor. Past and potential infrastructure bottlenecks at the border, for example in transportation (USITC, 1991, p. 5.6), are assumed to create another incentive to locate production in non-border areas.

The data presented is based upon élite interviews and two surveys of maquila managers and workers completed in 1989 by the author, R. Haywood and R. Bolin in the interior and along the border of the state of Sonora, with the special assistance of the Nogales Maquila Managers Association and the Instituto Tecnologico de Monterrey at Hermosillo. The composition of the Sonoran maquila sample by sector of activity approximates that of Mexican maquilas as a whole, with a predominance of the electric/electronics sector and a marginalization of the low-technology apparel sector – no more than 10 per cent of total maquila activity at the beginning of the 1990s.

17.2 DIFFERENCES AMONG MAQUILA WORKERS: BORDER vs INTERIOR

The following results are drawn from direct questionnaires and interviews conducted with maquila workers from eleven separate maquila firms in five different Sonoran cities. Workers were from five border firms at Nogales and six interior firms at Hermosillo, Guaymas, Empalme and Cd. Obregon. The questions were designed to be as freely answerable as possible, and workers were told that their answers would be kept anonymous. Independent interviewers were from among the faculty and students at the Hermosillo campus of the Instituto Tecnologico de Monterrey.

17.2.1 Workforce Characteristics and 'Maquila-Hopping'

The Nogales sample had more quality-control workers (36 per cent of the total workforce) than the interior sample (14 per cent), possibly indicating higher quality control needs at Nogales due to higher technology. Nogales workers average 2.2 years with their firm (with a maximum of 10 years), whereas the workers in the interior averaged 0.13 years (with 1.1 years maximum). As an indicator of the higher level of turnover and maturity at the border, 64 per cent of the Nogales sample had worked in other maquilas compared with only 21 per cent in the interior. Of those workers with previous maquila experience, 19 per cent in Nogales had had still another maquila job compared with 17 per cent in the interior.

These percentages suggest that rather than leaving the maquiladora industry entirely, as critics of the program and of NAFTA have assumed, a significant number of border workers in fact leave one maquila employer only to find work with another. Turnover due to 'maquila-hopping' is explained by the differences in fringe benefits that can exist from maquila to maquila, with U.S. and other foreign-based firms using benefits to attract and then retain workers. In maquila locations such as Ciudad Juarez and Chihuahua City, firms have attempted to standardize their menu of fringe benefits, hoping to discourage their employees from leaving for benefits that may be only marginally better elsewhere (see Chrispin, 1990, p. 76).

17.2.2 Workers' Attitudes towards the Maquila Program

When asked to compare their present work with former jobs, 40 per cent of the Nogales sample found the work better, 8 per cent equal, 16 per cent worse, and 36 per cent had no opinion. In the interior, 48 per cent found the work to be better than before, 10 per cent equal, none found the work to be worse and 41 per cent had no opinion. If given the opportunity, 36 per cent of the Nogales sample would change companies against only 10 per cent in the interior. This is an additional indicator of potential labor instability on the border, which is useful information for firms seeking to locate Mexico in the event of NAFTA taking effect. Nogales workers preferred the U.S.-owned to the Mexican-owned maquila by a ratio of three to one when the question applied, whereas the workers in the interior were equally divided over their preference.

In response to the question: 'Is your standard of living better or worse than two years ago?', no less than 88 per cent of the Nogales workers surveyed replied that it was better or equal (48 per cent better plus 40 per cent equal). Only 8 per cent found their standard of living to be worse; 4 per cent had no answer. In the interior 76 per cent of the workers found their standard of living to be better or equal (66 per cent better plus 10 per cent equal), 17 per cent were worse off and 7 per cent had no answer. When asked about their expected standard of living in four years' time, responses were again highly favorable, even though lower than with respect to the recent past. In Nogales 60 per cent expected that their future standard would be better or equal to their present standard (24 per cent better plus 36 per cent equal). In the interior 62 per cent believed the future would be better or equal (48 per cent better plus 14 per cent equal). In Nogales 20 per cent felt they would be worse off in four years' time, compared with 14 per cent in the interior.

In Nogales, 84 per cent of the workers felt the maquila program had benefited them personally, against 72 per cent in the interior. Nogales workers were generally more positive about the benefits to themselves and to Mexico of the maquilas than were workers in the interior.

An explanation for these findings concerning the perceived benefits and improved standard of living due to the maquiladora industry lies partly in similar results from a joint ILO/UNCTC study on the effects of EPZs. That study noted that 'the salaries of maquiladora workers tend to be considerably higher than those of workers in other Mexican industries' (1988, p. 77). In the same manner Salinas (1986, pp. 25–6) found that 'the maquila worker fares favorably with other Mexican workers in the manufacturing industry. Thus, it is difficult to conclude that the maquila worker is an 'exploited' class among other Mexican workers'. The above results all take as their central point of reference the Mexican economy rather than the U.S. economy and its wage levels for unskilled workers. 'The Mexican worker is not trying to subsist in the U.S. economy but in the Mexican economy' (Salinas, 1986, p. 25).

Other explanations for the positive impact of the maquiladora industry on Mexican labor, as affirmed by its workers, has been provided by a researcher from the U.S. Department of Labor's Bureau of International Labor Affairs. G. Schoepfle found that U.S.- and Japanese-owned maquilas located in industrial parks had high standards of safety and health relative to the 'much less desirable' conditions in domestically owned firms and smaller assembly firms outside the industrial parks that maquiladoras represent (1989, pp.

25–6, 29). Even C. Darman, a critic of the Nogales maquila program and opponent of NAFTA on the basis of environmental issues and community health outside the factories, has stated that 'in the course of just a few years the maquiladoras brought the industrial revolution and all that goes with it to a region that was completely unprepared for it. In the short run this is a positive development because a job is better than no job and because living conditions where these people came from were also terrible' (Suro, 1991). Managers' opinions on environmental and infrastructure problems are analyzed in section 7.3 below.

17.2.3 Benefits: Border vs Interior Sonora

In response to the question: 'Apart from salary, do you receive other benefits from this firm?', 80 per cent of the Nogales workers surveyed answered in the affirmative; 55 per cent of these cited bonus and loan programs as the benefits they most appreciated. Another 25 per cent cited food, 5 per cent transport and 15 per cent named various other benefits, such as time off and special events such as picnics. Only 8 per cent of the Nogales sample said they received no benefits and 12 per cent gave no answer. In sharp contrast with the border, only 48 per cent of the workers in the interior answered yes, against 38 per cent no, indicating a greater absence of benefits. The competitive labor market on the border explains the greater benefits. Nogales maquilas provide lunch as a benefit, for example, but not the interior. Such differences have helped incite foreign firms to choose an interior investment location, everything else being equal, and can be expected to do the same in the event of NAFTA being implemented.

17.2.4 Variable Education Levels: The Border vs the Interior

The average number of years of schooling was 8.7 at Nogales compared with 10.6 years in the interior. The lower level of education of the Nogales worker sample relative to the interior can be interpreted as another indicator of the saturation of the unskilled labor force along the border. More highly educated persons are available in the interior, in what constitutes yet another incentive for firms to locate production in the interior given a freetrade agreement. To successfully compete with the Mexican interior as a plant location once NAFTA is in place, a significant upgrading of education and skill levels at the border appears necessary. The Flagstaff Institute has recommended the 'deliberate importing' of engineers, technicians and other skilled workers from cities on the U.S. side of the border, in addition to adult education, night classes and specialized training (Haywood and Bolin, 1989, pp. 59–60).

17.2.5 Industry Impact on Labor Migration

Three quarters of the workers reporting in the Nogales sample were born in the interior and the others in Nogales, indicating the role of Nogales as a pole of attraction for labor in the 1980s. Only one worker in the interior sample was from Nogales.

Proponents of NAFTA have argued that increased trade and foreign investment in Mexico will slow the flow of Mexican labor to the United States. The results for Nogales suggest that the maquiladora industry has indeed attracted and retained workers who have spent a significant number of years there. The industry represents a pole of development, especially with the introduction of increasingly complex technology (see Chrispin, 1990, pp. 83–5). It does not appear to be serving as a temporary halt on the road from the Mexican interior to the United States, as has been alleged by its critics. Similar results have been found by researchers at other investment locations along the border, for example Brannon and Lucker (1988, p. 4) for the Ciudad Juarez area. The maquilas in the interior of Sonora are settling into greenfield sites with even more stable labor environments.

The survey of managers found that women constituted an average of 49 per cent of line workers in border maquilas, indicating an increased presence of male workers. This finding is consistent with the general trend in the maquiladora industry (Warner, 1990, pp. 187–8; Chrispin, 1990, p. 74).

17.2.6 Workers' Life-Styles and their Impact on US Border Economies and Retail Firms

Nogales reported a lower average household size, but more maquila workers per household, compared with households in the interior. More than half of the workers in both samples share their bedrooms. The 2.5 maquila workers per household in Nogales – precisely the same result as that found for the border town of Ciudad Juarez by Brannon and Lucker (1988, p.52) – suggests more-crowded living conditions, with single individuals sharing housing with other individuals. Workers in the interior, on the contrary, tend to live at home with their families. All workers in both samples had running water, electricity, radio and television. For several workers in both samples, however, the water was not always potable, and the house did not have a sewer. Only 16 per cent of homes in Nogales had a telephone, in comparison with over 27 per cent in the interior. In Nogales 60 per cent of households had a car, but that car was never owned by the maquila worker. In the interior 79 per cent of the workers surveyed lived in a household with a car, but only 7 per cent stated that the car was their own. One-third of the Nogales workers sample paid rent as compared with 14 per cent in the interior.

In response to the question: 'Do you shop in the U.S.?', 56 per cent of the Nogales workers were found to shop frequently in the United States: 40 per cent on a weekly basis plus 16 per cent every two weeks on the average. In sharp contrast, only 20 per cent of the interior workers surveyed shopped in the United States, and then very irregularly: 10 per cent rarely, 7 per cent annually and 3 per cent on a monthly basis.

Most of the weekly shoppers from Nogales were found to spend $20 per visit on average. The biweekly shoppers averaged $20 per visit. These results indicate that despite the falling buying power of the Mexican peso in U.S. dollar terms in the 1980s, a significant share of the incomes of maquila workers along the border has been devoted to the consumption of U.S. goods and services. The average propensity of the Nogales maquila worker to import U.S. consumer goods would be 25 per cent of income, hence the favorable impact of the maquiladora industry on U.S. retail sales and employment in the border region. This result is higher than the 5–15 per cent cited by Clement and Jenner (1987, p. 74) for the Tijuana area. In general, several U.S. border cities owe 20–30 per cent of their growth to the multiplier effect of nearby maquilas (Fatemi, 1990, p. 8). In El Paso, Texas, Mexican maquilas were found to be responsible for one in five jobs (Patrick, 1990, p. 60).

With NAFTA, U.S. retailers would be able to penetrate into the Mexican interior and reduce the observed drop in the propensity of maquila workers to consume U.S. goods as the distance from the border increases. The USITC (1991) has forecast the move of large U.S. retailers to the interior, as Wal-Mart, the largest U.S.-based retailer, did in 1991 upon forming a joint venture with Cifra S.A. Smaller retailers are expected to remain at the border.

17.3 MAQUILA MANAGEMENT ISSUES AND OPINION

The following results are based on a separate survey which included interviews with maquila plant managers in Sonoran border cities and the interior; analysis of questionnaires completed by maquila managers located in four border cities (Nogales, Agua Prieta, San Luis Rio Colorado and Naco); and interviews with U.S. and Sonoran developers, investors and government authorities associated with the maquiladora industry.

17.3.1 Investment Location Factors

Managers were asked to assess the relative importance of numerous investment-location factors by using the following categories: (0) indicates the factor is not an important influence on plant location decisions; (1) the factor is desirable; (2) the factor is fairly important; (3) the factor is extremely important. The most important factors were considered to be ease of customs procedures for goods (2.89), reliable electric power (2.74), a stable labor environment (2.58), a reliable water supply (2.58), availability of skilled labor (2.53), quality of housing available (2.53) and proximity to the United States (2.47).

Factors found to be relatively less important included the availability of truck service (the means of transport for over 90 per cent of the inputs of the survey's respondents) (2.42), support of local government officials (2.42), availability of middle managers (2.42), the political stability of Mexico (2.37) and low wage rates (2.37). The factors judged to be least important were the ability to work night shifts (1.79), differences in environmental regulations between the United States and Mexico (1.63), travel time to parent plants (1.53), the investment of similar companies in the area (1.39), financial incentives offered by the government (1.37), and the availability of air freight (1.21) and rail transport (0.61), which together are used by less than 10 per cent of the firms surveyed.

This evidence suggests that low wage rates – while deemed important – were not judged to be the primary factor influencing investment location, as critics of both the maquiladora industry and NAFTA have assumed. This survey result is consistent with studies such as that of the International Labour Organisation and United Nations Centre on Transnational Corporations, which found that 'low wages alone, or even total labor costs, are seldom, if ever, the only determinant in the decision to invest in an export processing zone . . . very low wages may induce a foreign firm to consider seriously the possibility of investing in one country rather than another, but ultimately the decision is motivated by such factors as the quality of the work force, its educational level, its willingness to work and more generally its potential productivity. At a deeper level, its decision to invest is very strongly influenced by the firm's perception of the host country's political stability, and more generally its overall social and political climate' (ILO/UNCTC, 1988, p. 90). Similarly Drucker (1988 and in Bolin, 1988, p. 29) has observed that low wages alone do not guide location decision-making by firms engaged in production sharing. Productivity will be increasingly important, as it was for the managers surveyed. Proximity to the final market is also very critical: low wages as well as high market-sensitivity explain, for example, Sony's integrated Tijuana/San Diego operations in consumer electronics (Bolin, 1988, pp. 65–6).

Proximity to the United States as a critical investment location factor concerns not only the target market, but also potentially lower transport costs and a better opportunity for control over day-to-day manufacturing operations. The Clement and Jenner survey of U.S.-based firms with maquila operations in the Tijuana area found the transport and control factors to be two of the five most important reasons why Mexico is more attractive as an investment site than other locations such as Asia (1987, pp. 66–8). The quality of life in southern California, from where U.S.-born maquila managers have commuted, and the availability and cost of labor were the other key reasons.

The emphasis by maquila management on the reliability of utilities and the quality of infrastructure in general is indicative of growing pains at an investment location experiencing rapid expansion, especially when the location is in a developing country. At greenfield investment locations, infrastructure bottlenecks in transport, energy generation and telecommunications are factors that typically threaten business expansion.

The very low importance of differences in environmental regulations between the United States and Mexico as a location factor can be explained by the efforts of Sonoran developers and authorities to avoid attracting polluting industries, and by the sizable presence in the sample of the relatively 'clean' electronic/electrical and apparel industries. This result is in contrast with a GAO study which specifically focused upon Los Angeles firms in the furniture industry that were relocating to Mexico. Stringent air-pollution emmission-control standards for paint coatings and solvents were cited as a reason for the relocation by 78 per cent of the firms surveyed (U.S. GAO, 1991, p. 4).

17.3.2 Productivity and Turnover

The labor productivity of the maquila firms completing the questionnaire was found to average 80 per cent of U.S. labor productivity. Thirty-seven per cent of the maquila managers reported labor productivity of 100 per cent or more. Twenty-one per cent of the plants reported 50 per cent or less of the U.S. parent's labor productivity.

Managers were asked questions in the form of statements to which they could either strongly agree ($+2$), agree ($+1$), express no opinion (0), disagree (-1), or strongly disagree (-2). Results were separated into Nogales managers' opinions and managers' opinions in the other border locations. Nogales managers more generally agreed with the statement 'The labor force here is very productive' than did the other managers (1.11/0.30). Managers disagreed with the statement 'Labor relations in Sonora are unstable' ($-1.00/-0.3$).

For Mexico in general, maquila workers are considered to be easily trainable, a reliable source of labor with productivity as high as in the United States and quality control that is often superior (see Chrispin, 1990, p. 81; George, 1990, pp. 224, 229). Kreye, Heinrichs and Frobel (1987, p. 18) have concluded that productivity in the plants of developing country EPZs – of which Mexico's maquiladoras are a prime example – usually matches that in industrial countries. The claim that low-wage countries have low productivity is unwarranted as far as foreign-owned plants are concerned. For domestically owned firms, however, low labor costs might reflect low productivity and low quality (Dornbusch, 1991).

In spite of the overall positive assessment of the maquilas' labor productivity, turnover inhibits future productivity growth and was ranked by managers as the worst production problem. Turnover in the border region will discourage high-tech industry from locating there in the event of NAFTA being implemented, to the benefit of the interior.

The results contradict the viewpoint of Tiano (1987), that maquila management encourages high turnover in order to maintain profits. The average level of acceptable turnover found by the Sonora survey was similar to the rate of less than 7 per cent per month found by Lucker and Alvarez (1984). With the help of fringe benefits, turnover for basic assembly workers is assumed to have fallen throughout the industry to levels comparable with those in the United States for similar occupations (Chrispin, 1990).

Key causes of turnover in Sonora were found to be inadequate infrastructure – poor housing, transport to workplaces and sewage – as well as low compensation and thus 'maquila-hopping' in search of better fringe benefits. Industry researchers and practitioners elsewhere in Mexico also include the monotonous, repetitive nature of the tasks performed, the departure of newcomers to the world of manufacturing, insensitive treatment of employees by management, the entry of less qualified people into the industry as the demand for labor rises, and the reluctance of foreign firms to raise wages and thus risk creating problems with local firms unable to follow suit (Bolin, 1988, pp. 33–5; Clement and Jenner, 1987; Warner, 1990; Williams, 1988).

Compensation differentials and a shift by maquila plants to Mexico's interior are seen as two key solutions to turnover in Sonora. A standard menu of fringe benefits, infrastructure improvements and better training and counseling for new employees have also been proposed and implemented elsewhere (see Rivas and Sada, 1988, pp. 69–73).

17.3.3 Automation

One-half of the firms surveyed did not adapt their production process when opening a maquila operation; the other half did make various modifications. Managers along the border agreed that their products would require less labor per unit in the future (0.44 Nogales/0.30 other border locations). They agreed that an increase in automation in their maquilas was planned (0.56/0.60). But at the same time both samples disagreed with the statement that automation would reduce the need for their factories (−0.56/−0.30). Finally, the managers rather strongly agreed that the production processes in their plants were becoming more specialized (1.56/1.30).

These results suggest that rather than constituting an alternative to the maquila, automation will accompany their future development, increasing the level of value added per maquila worker. Research into the arrival of a second generation of more technologically sophisticated, capital-intensive maquilas, operated by a new generation of highly competent Mexican engineers, technicians and middle managers, has reached similar conclusions (Barrio, 1988; Bermudez, 1990; Chrispin, 1990). CAD/CAM used by Mexican technicians is already common. The probability that this new wave of maquilas will be footloose – an allegation frequently made by critics of EPZs – appears to be lower than for the earlier generation of maquilas, and especially in the event of NAFTA being implemented.

17.3.4 Linkages, Sourcing, Destinations and NAFTA Rules of Origin

In addition to providing hard currency to service foreign debt and relieving high unemployment, one of Mexico's key goals in seeking to attract investors to its maquiladoras has been to forge backward linkages between Mexican firms supplying inputs and the export-oriented maquilas. The 1989 decree on maquiladoras underscored the important role assigned to linkages in order to enhance the competitiveness of Mexican-owned firms. (The 1989 decree required new maquilas to achieve content of 2 per cent in their first year of operation, 3 per cent in the second and 4 per cent in subsequent years. Mexican firms were to be exempt from valued-added taxes on production sold to maquilas.) With respect to the near future, however, very few managers in Sonora indicated that they expected to use more Mexican inputs (0.11/0.00). They largely agreed that they would like to see more local sources of supply (1.56/1.10). The low quality of local material was found to be a cause of this absence of linkages in Nogales (0.44/0.00). The insufficient quantity of production by Mexican suppliers was considered to be a factor in cities other than Nogales, but not in Nogales itself (−0.44/0.60). To promote increased linkages with Mexican industrial firms the managers surveyed indicated a willingness to offer technical assistance to a local supplier (0.22/0.50), but not financial assistance (−0.11, −0.30).

Minimal local content is the norm for maquilas throughout Mexico – only 1.5 per cent in 1987 – and contrasts sharply with an integration of local parts as high as 40 per cent in many Asian EPZs. Taiwan's domestic firms on average supply 25 per cent of all EPZ inputs on the average (Wang, 1990, p. 24). Mexican inputs have principally been limited to raw materials such as wood and food, followed by packaging. The Ciudad Juarez maquiladora developer S. Bermudez has targeted such manufactured inputs used by maquilas as metal casting and molded parts for future sourcing in Mexico rather than in the United States and other foreign countries. The developer's fundamental objective is to increase the competitiveness of Mexican industry internationally by forging links between Mexican parts suppliers and nearby maquiladoras (Bermudez, 1990a, p. 121). Everything else being equal, successful Mexican sourcing of maquila inputs would have a negative impact on current non-Mexican sourcing, which is primarily of U.S. origin.

For 79 per cent of the maquilas surveyed in Sonora, 100 per cent of output left Sonora; 95 per cent of this total output was shipped, usually by truck, to the foreign parent or client firm. On average, maquila output was shipped to only 1.9 locations. For 74 per cent of the maquilas one of these destinations was at the same time the primary supplier, thus indicating the essential function of the average maquila as a cost center at the beginning of the 1990s.

NAFTA would end the traditional refusal of Mexican business and government to allow maquila products to compete with local products, an anti-competition policy distortion that is in sharp contrast with the successful Korean development model, where EPZ-designated firms have been able to produce for the Korean domestic market as well as for export (see Grunwald in Bolin, 1983, p. 95; Schoepfle, 1990). The Sonora survey of managers found that approximately two-thirds of the firms would expect a market for their goods in Mexico.

17.4.5 Maquila Workers' Welfare: Its Impact on Firms' Performance

Managers were found to disagree with the statement 'Health and safety standards in this plant are below U.S. standards' (−0.78/− 0.10). This response concerning conditions inside maquila plants in Sonora is consistent with investigations by the U.S. Department of Labor's Bureau of International Labor Affairs. After studying labor standards in Mexican and Caribbean EPZs, G. Schoepfle (1989, pp. 25–6) concluded that 'in terms of the physical workplace setting, the Mexican maquiladoras − in particular those owned by U.S. or Japanese multinationals and located in industrial parks − probably offer the best occupational safety and health conditions. They are well-lighted and ventilated, and similar to comparable plants located in the United States'.

Schoepfle found Mexican legislation on occupational safety and health to be 'relatively advanced', particularly in comparison with such Carribean nations as the Dominican Republic, Haiti, Costa Rica and Jamaica. But the monitoring and enforcement of existing laws can be 'woefully inadequate' (1989, p. 25).

Concerning workers' welfare outside the maquila plant, Nogales managers agreed that solid-waste disposal was a problem (1.00). In locations other than Nogales there was neither agreement nor disagreement (0.00). Both groups of managers did disagree, however, with the statement 'The facilities for disposal of hazardous waste material are adequate' $(-0.78/-0.40)$. As of 1991 the U.S. State Department and Mexico are investing $16 million in Nogales to expand its international sewage treatment plant (Reilly, 1991).

A critical concern of maquila managers with respect to workers' welfare outside the plants in the border cities involved inadequate housing, transportation and utilities. The state of Sonora was perceived as not helping meet workers' needs in spite of available resources. The 5 per cent wage tax levied on maquila employers, for example, in order to build and improve housing for workers, allegedly produced few visible results in border communities. Some maquila managers saw inadequate living standards, whenever they existed, as having a negative impact on maquila performance and quality, in addition to fueling criticism of both the industry and Mexican goverment authorities.

18 Trade Relations between the United States and Mexico

Nancy A. Wainwright

18.1 INTRODUCTION

The two geographically closest trading partners to the United States are Canada and Mexico. On 1 January 1989 the United States and Canada entered into a free trade agreement which will phase in the free flow of goods and services between the two countries over the next ten years. Free trade between Mexico and the United States progressed significantly when Mexico signed the General Agreement in Tariffs and Trade (GATT) in 1986. The culmination of this liberalization came on 12 August 1992 when the North American Free Trade Agreement (NAFTA) was concluded, and which could go into effect on 1 January 1994. The agreement will eliminate tariffs between Canada, the United States and Mexico over fifteen years.

Particular industries are dealing with the advantages and disadvantages of expanding free trade with Mexico. This chapter will examine the needs of the fruit and vegetable industry. The issues of product safety, market orders and distress pricing were identified at hearings prior to fast-track approval by the United States Congress. Specifically, since the conclusion of NAFTA the three countries have agreed to strive for a domestic price-support policy that does not distort trade in fruit and vegetables.

With the 23 May 1991 House of Representatives fast-track approval and the 24 May 1991 Senate fast-track approval, President Bush was given the power to negotiate a free trade agreement with Mexico.

The amount of trade between the United States and Mexico is valued at between $33 billion and $42 billion, which makes Mexico the United States' third-largest trading partner behind Canada and Japan. Trade between the two countries creates jobs in both countries and is essential to the financial health of both countries. Exports from Mexico provided Mexico with the money necessary to finance payment of its foreign debt. Exports to Mexico help the United States' balance of trade.

Mexico has a young population, with 64 per cent of the population under the age of 24 and 43 per cent under the age of 14. This type of population base provides the United States with the young workers it lacks as well as with a growing consumer market. In determining whether or not there is a consumer market, one of the factors to be considered is the education of the population. Mexico's population is 80 per cent literate (Center for Strategic and International Studies, 1989). Mexico has 85 million potential consumers, however at the moment only 10 million are economically able to be consumers – 75 million live from hand to mouth and are only able to provide themselves with food and shelter.

18.2 TRADE HISTORY

In the past Mexico had a formal protectionist policy, which included the use of closed-door import substitution, prior licensing and ad valorem tariffs (Verbit, 1969). Prior to 1982 100 per cent of the value of Mexican imports were subject to prior licensing, but in 1992 less than 20 per cent remained subject to this type of licensing. With the signing of GATT the Mexican government eliminated the licensing requirements on all but 232 of 1000 items on the tariff list (Weintraub, 1990). Import licensing required the permission of Mexican customs for goods to enter the domestic market, which effectively resulted in complete protection for Mexican products when a license was refused (Stare, 1987). Mexico used as its tariff an ad valorem tax, that is, a percentage increase added to value, ranging from zero to 20 per cent. In 1987 it imposed a general import tax of 5 per cent, but that has now been removed.

The nontariff barriers used by Mexico are selective subsidies, export performance and local or domestic content requirements (Dacher, 1990). Selective subsidies are granted to certain products, resulting in the domestic products being lower in price than equivalent imported goods. Export performance requirements are used to enable the development of intermediate-goods production in Mexico by forcing trading partners to accept some Mexican exports, and Mexican trade policy went one step further by seeking to balance imports with exports in specific sectors.

The nontariff barriers used by the United States to protect domestic industries against Mexican imports have included health regulations on fresh vegetables and meats, border-crossing and customs regulations, the lack of bridges between the two countries in the lower Rio Grande Valley, and contingent protection (Weintraub, 1990).

18.3 CHANGES IN THE TRADE POLICY

Initial changes in the trade policy between the United States and Mexico began on 23 April 1985 with the 'Understanding Between the United States and Mexico Regarding Subsidies and Countervailing Duties'. This agreement granted most-favored nation status (MFN) to Mexico. Prior to entering into this agreement the United States had twenty-eight cases against Mexico for dumping, more than against any other nation (*George Washington Journal of International Law and Economics*, 1984). In 1985 the United States extended the general system of preferences to Mexico that were established in the Trade Act of 1974.

The most significant step to date in the development of free trade between the United States and Mexico was Mexico's accession to GATT on 24 July 1986, making Mexico the 92nd contracting party (U.S. Department of Commerce, 1989). Mexico also signed the GATT Code on Standards, Licensing, Customs Valuation and Subsidies on 21 July 1987 (U.S. Department of Commerce, 1988) With Mexico's accession to GATT, the trade policies of the two countries became more closely aligned.

This progress toward free trade is evidenced by the United States and Mexico entering into bilateral agreements. The first was signed in November 1987 and was entitled 'Understanding Concerning a Framework for Trade and Investment' (U.S. Congress, 1989). This established a mechanism to facilitate consultation between the two parties about matters concerning bilateral trade and investment relations. Subsequent agreements have included the December 1987 bilateral agreement on textiles, steel and alcoholic beverages, a formal agreement on textiles signed on 13 February 1988, and the 1989 agreement entitled 'Understanding Regarding Trade and Investment Facilitate Talks'.

18.4 FREE TRADE BETWEEN THE UNITED STATES AND MEXICO

The study of free trade between the United States and Mexico began with hearings before the Subcommittee on Western Hemisphere Affairs of the Committee on Foreign Affairs of the House of Representatives on 7 June 1989, in which the committee endorsed the long-term goal of free trade with Mexico. The committee was in favor of such a move in order to rival the European Common Market, because of the growing interdependence of the two countries, and because Mexico is geographically close, rich in natural resources, has a large working population, is politically stable, has large markets, and is developing economically (Overview of United States–Mexico Relation, 1989).

The United States International Trade Commission (USITC) began a two-phase investigation of Mexico's recent trade and investment reform on 8 November 1989 following a request by the House Committee on Ways and Means on 18 October 1989. Phase one of the investigation will include a comprehensive review of the recent trade and investment liberalization measures undertaken by Mexico and, where feasible, a description of the implications for U.S. exporters and investors. Phase two will study free trade areas, an enhanced dispute settlement mechanism, possible sectorial approaches (Bilateral Trade Agreements, 1989), the framework of understanding and other options for bilateral trade agreements (Code of Federal Register, 1989).

On 14 and 28 June 1990 the Subcommittee on Trade of the Committee on Ways and Means of the House of Representatives held hearings on free trade with Mexico, and on 10 October 1990 the USITC began another investigation following a request on 28 September 1990 by the Committee on Ways and Means of the House of Representatives and the Committee on Finance of the United States Senate. Specifically, the USITC was requested to report on the following:

(1) an overview of recent events significantly influencing U.S.– Mexico economic relations, including a profile of Mexico's trade and investment patterns, (2) a summary of the likely impact of the proposed free trade agreement with Mexico on the U.S. economy in general, (3) a summary of the likely impact on major U.S. industries and other sectors, including agriculture that would be most affected by the proposed free trade agreement with Mexico, and (4) an indication of the regions in the United States that would be most affected by the proposed free trade agreement with Mexico and a summary of the nature of these effects. The USITC was also requested to study the impact of a three way interrelationship among the United States, Canada, and Mexico.(Code of Federal Register, 1990)

On 13 February 1991 the USITC itself initiated an investigation aimed at gathering the information necessary to provide the president with advice on the probable economic effects on U.S. industries and consumers of the removal of U.S. import duties on Mexican goods. The USITC will study the list of items submitted by the United States Trade Representative (USTR), which the USITC anticipates will be all dutiable products listed in column one of the Harmonized Tariff Schedule (Code of Federal Register, 1991).

The common concerns of Congress that were discussed or testified to in the above hearings were labor issues, involving the loss of jobs (Spokesman Review, 1991), safety and health issues in the workplace, child labor practices, and environmental issues on the use of pesticides (Cong. Rec., 1991).

18.5 FREE TRADE AND THE FRUIT AND VEGETABLE INDUSTRY

In addition to the discussions of labor and the environment, Congress was also presented with testimonies from fruit and vegetable growers, who produce crops such as avocados, tomatoes, citrus fruits, apples, broccoli, lettuce and brussel sprouts in states such as Texas, Arizona, California, Florida and Washington.

In February 1991 the subcommittee on Trade of the Committee on Ways and Means of the House of Representatives heard testimonies from various fruit and vegetable interest groups:

1. The Committee on Agriculture of Florida, Beardsley Farm, the Mumm Packing Co., the Western Growers Association of California and Arizona and the United Fresh Fruit and Vegetable Association were unhappy about the free trade agreement and requested that their industry receive exceptions in the agreement.
2. Furman Foods Inc., Crystal Valley Farms, Cherokee Products Co. and the Florida Citrus Industry accepted the idea of free trade but wanted to be specifically excluded from the agreement.
3. The American Dehydrated Onion and Garlic Association was opposed to the free trade agreement on the basis of food safety. The California Avocado Commission was also opposed to free trade because of food safety, specifically because of the seed weevil in the Mexican avocado.

4. The American Soybean Association was in favor of free trade because of an increased export market. The California Cling Peach Advisory Board, Pacific Northwest Horticulture, Sunkist Growers Blue Diamond Growers, and the National Corn Growers Association were also in favor of free trade because of the expanded export market.

5. The Florida Tomato Exchange and the National Potato Council were opposed to free trade.

A further review of the testimonies finds consistent references by these fruit and vegetable industries to Section 22 quotas, snapback tariff, import licenses and health standards. The two most organized submission were presented by The Canadian Horticultural Council and the United Fresh Fruit and Vegetable Association. The former summarized the potential problems arising from free trade with Mexico by referring to the problems enountered since the signing of the Canada–U.S. Free Trade Agreement in 1989. The issues they addressed were the U.S. market order system, the pesticide registration system, antidumping and countervailing duty, comparable border points, acceptance of certification and inspection procedures, dispute settlement mechanisms and safeguard mechanisms to assist industry against distressed pricing, such as snapback tariffs and surtaxes (Committee on Agriculture, 1991).

The United Fresh Fruit and Vegetable Association expressed concern by referring to the issues as the following nontariff issues: phytosanitary and residue standards, import licensing, dumping code, transportation, disparity of government regulations and marketing orders (Committee on Agriculture, 1991).

The difference between government regulations in the two countries can be demonstrated by greater restrictions on the use of pesticides in the United States and much stricter regulations on farm workers' safety, minimum wages and migrant housing. However it should be noted that farm laborers have traditionally been excluded from minimum wage standards in order to keep food prices at parity or above, and substandard migrant farm housing would seem to indicate that producers are not cutting into their profits to provide their migrant workers with adequate housing. Other differences include land ownership. However, Mexican landowners are restricted to 250-acre farms, which could offset the issue of land value in the United States.

The U.S. General Accounting Office (GAO) also prepared a report for the committee in which it identified the growing need by Mexico to import corn, beans and dairy and meat products to meet the needs of its consumers; the overlap in the harvest and marketing season of cantaloupes, watermelons, table grapes and asparagus; and the direct competition of tomatoes, cucumbers, bell peppers and strawberries.

As a result of the dialogue on the trade, in 1990 Mexico imposed import licenses on only 57 agricultural products and in May 1991 further products were removed from licensing restrictions. Import licenses still exist on corn, animal fats, milk and dairy products. Mexico will continue to have import licenses as long as U.S. policies on product safety are so detrimental to the Mexican producer – if Mexico has a crop infestation, the United States imposes a one-year ban even though subsequent products have tested safely. The GAO also cited as constraints on trade the long delays at border crossings, which is an especially serious problem for crops which perish over time (Committee on Agriculture, 1991), other health regulations and inadequate Mexican infrastructure.

An exporter of fruit and vegetables to Mexico would have to find an alternative market should import licenses be reintroduced. This problem would be eliminated by a free trade agreement or at least addressed by a dispute-resolving mechanism or a damages agreement. The products which would suffer least from a reintroduction of import licensing are those for which production is not being increased on the basis of greater access to the Mexican market, such as apples, the growers of which are instead seeking the best price possible.

Producers of canned fruits and vegetables would see Mexico as a good alternative market because delayed border crossings and the poor infrastructure in Mexico would have no effect on their products. With items such as milk, however, the inability to reach the market, even a market that is so close, would present problems for both importers and exporters. Perishable goods cannot be held at the border while their right of entry is determined.

The food safety issue, like border crossing, needs a prompt-resolution mechanism. If a particular disease or pest is a long-standing problem in Mexico, then the product subject to it should be denied entry until such time as the pest or disease is eradicated. If a particular chemical affects current growth only, leaving future crops uncontaminated by that herbicide or pesticide, then the restriction or those crops needs to be removed promptly. The issue of food safety seems to be one of providing inspectors to respond to particular situations rather than one of Mexico seeking to violate U.S. regulations. A producer of fruit and vegetables in Mexico does not want to see his entire crop, his livelihood, stopped at the border and denied entry to the United States because of the use of proscribed herbicides or pesticides.

Potentially dangerous herbicides or pesticides also is an issue in the United States, as well as in Mexico, in that when these chemicals are banned in the United States they may be exported to foreign countries by U.S. firms. The demands for legislation in the United States would be strengthened by the argument that these potentially dangerous chemicals should not be exported to Canada and Mexico, who are exporting food into the United States (Satchell, 1991).

18.6 MARKET ORDER IN THE FRUIT AND VEGETABLE INDUSTRY

All three parties involved in the North America Free Trade Agreement want to protect their agricultural industry with stable regulations and distress pricing to prevent market collapse. The free trade agreement between the United States and Canada has already highlighted the market order system in the United States as a suitable method of regulating the fruit and vegetable industry among the three countries.

The Agricultural Act of 1933 Section 22 allows for the importation of several basic commodities – those whose domestic counterparts are eligible for price support – to be limited by quotas and fees. The Agricultural Adjustment Act of 1937 provides for a market plan as opposed to direct price setting for eligible vegetables, fresh and dried fruits and tree nuts. Both of these acts need to be addressed in future NAFTA negotiations.

The purpose of the Agricultural Adjustment Act of 1937 is to 'maintain such orderly marketing conditions for agricultural commodities in interstate commerce as will establish commodity prices at a level which will give them a purchasing power equivalent to such power in the "base" period which was fixed as between 1909 and 1914'. The statute authorizes the creation of committees of shippers on a pro rata basis to regulate the flow of agricultural products so as to stabilize the market and lessen fluctuations in price. Market orders may regulate shipments in interstate and foreign commerce by limiting the amount that may be shipped, by the control and disposal of 'surplus', which may be affecting price parity, by prior inspection and certification of grade requirements, and by limiting the annual marketable quantity each grower is permitted to produce. These market orders may also control grades, containers and anti-consignment regulations. The market-order system which exists in the United States could provide a level playing field for the three countries' producers.

18.7 CONCLUSIONS

A review of the literature results in the conclusion that a critical issue is how to prevent a market collapse in a particular agricultural product, an issue that the United States has been addressing since the Great Depression and continues to address in the free trade and GATT roundtable discussions. A North American free trade agreement would facilitate improvements in such nontariff issues as the more uniform use of dangerous pesticides and herbicides among the three participating countries and an improvement in regulation and certification at border crossings. Mexico would benefit from increased expenditure on its transportation infrastructure. The use of a system similar to the market order of the United States could help alleviate the problem of pricing among the three countries. This improvement in trade relations would benefit all three countries whether or not a free trade agreement is negotiated and whether or not GATT is able to resolve the issue of subsidized agriculture.

19 U.S.–Mexican Trade Opportunities: Toward the Development of a Globally Competitive North American Apparel Industry

Sandra Forsythe, Mary E. Barry and
Carol Warfield

19.1 INTRODUCTION

As a result of relaxations on trade restrictions, economic reform in
Mexico and the pending U.S.–Mexican free trade agreement, there will
be increased opportunities for cooperation between U.S. and Mexican
apparel manufacturers. Some see Mexico's entry into NAFTA as the
first step toward a Western hemisphere-wide free trade agreement.
Such a concept envisions a Western hemisphere trading bloc that could
dominate global trade. Complementing the present Canada–U.S. FTA
with the unique advantages offered by Mexico, a globally competitive
apparel-manufacturing industry is possible. Indeed, the combination of
labor and market power could result in an apparel industry that will
compete effectively in a global market. However, it is essential for both
U.S. and Mexican apparel producers to understand and act on these
opportunities. This chapter will address the market opportunities for
increased apparel trade between the United States and Mexico and the
potential for the development of a globally competitive apparel
industry.

The textile and apparel industry is becoming increasingly globalized.
Intensified competition worldwide has led to major shifts in production
to low-wage countries. Producers in western countries therefore face
slower market growth. Despite the existence of quotas under MFA, the
U.S. apparel industry has seen its market share deteriorate.

The global market share of U.S. firms has remained small, largely because U.S. apparel producers have not been able to respond aggressively to consumers' needs in the international marketplace. There have been a number of recent developments which suggest that Mexico has tremendous potential as a source of economic labor *and* as a foreign market for U.S. apparel. More importantly, by complementing the present Canada–U.S. FTA with the unique advantages offered by Mexico, the development of a globally competitive apparel manufacturing industry is possible.

As a result of relaxations on trade restrictions and economic reform in Mexico, many U.S. apparel manufacturers have moved to Mexico in pursuit of low-cost labor. The Caribbean Basin Initiative and Mexico now account for 17 per cent of apparel imports, up from 8 per cent in 1987. Mexico has a growing population and the median age of Mexican citizens is only 19. Almost 80 per cent of the population is literate, thereby providing a valuable labor pool. Not as well-publicized, however, is the Mexican market as a target for U.S. apparel firms seeking an export market.

19.2 MANUFACTURING AND MARKETING POTENTIAL IN MEXICO

The proposed North American Free Trade Agreement will provide excellent opportunities for U.S. apparel manufacturers to export to Mexico, particularly to the more affluent sector of the Mexican market with almost 30 million consumers. Since the early 1980s Mexico has surpassed every country in Europe in its importance as a manufacturing trading partner for the United States. Exports to Mexico nearly doubled between 1986 and 1990, making Mexico the fastest growing major market for U.S. exports (Fogarasi, Mohn and Subrin, 1991). Although it is not certain how much of this growth is due to the 807 and 807A provision, whereby garments are cut in the United States, shipped to Mexico for assembly, then shipped back to the United States, the potential for textile and apparel firms to export to Mexico is significant. According to the executive director of the U.S. Apparel Industry Council, 'Mexico is the best export market we have for U.S. apparel' (Marsh, 1991, p. 48). The total Mexican market for apparel, footwear and accessories is expected to grow at an average annual rate of 15 per cent during the period 1991 and 1994, achieving over $4148 million by 1994. A growing number of U.S. retailers and marketers are already targeting Mexican middle class consumers (Baker and Walker, 1991).

Current economic conditions make Mexico a prime target for increased cooperation with U.S. textile and apparel manufacturers. Mexico's GNP was $201 billion in 1991, when its GDP grew at a rate of 3.9 per cent last year (March, 1991). In addition tax rates have dropped following stricter enforcement of revenue collection. Furthermore, according to a report by the Federal Reserve Bank of St. Louis, Mexico's money supply and its consumer prices have risen dramatically. Tariffs have been reduced, inflation lowered, numerous government-owned companies have been privatized and confidence in the economy has been restored (Spiers, 1991). The liberalization of the investment regime is particularly noteworthy. One hundred per cent foreign ownership is now permitted in about three-quarters of the economic sectors. Moreover the procedures and regulations for obtaining permission to own a business in Mexico have been simplified.

U.S. apparel manufacturers enjoy a number of advantages over other countries in competing in the Mexican market. Mexico's proximity to the United States permits fast delivery, lower transportation costs, lower taxes and easily accessible servicing and technical assistance (U.S. Department of Commerce, 1990). Due to cable TV, satellite and other communications, Mexico is an attractive target for the absorption of U.S. culture and products. U.S. apparel brand names are well known in Mexico whereas there are no strong Mexican apparel brand names. U.S. brands have a reputation for quality, but perhaps most importantly, Mexicans are faithful to the U.S. brands which symbolize the North American experience for them (Baker and Walker, 1991).

There has been a noticeable increase in the sophistication of Mexican marketing techniques and distribution channels. Many Mexican retailers have adopted American-style marketing techniques such as self-service merchandising, and large shopping centers are being developed in the suburbs of major Mexican cities. U.S. apparel firms enjoy excellent brand-name recognition among middle and upmarket Mexican consumers for whom quality and styling are especially important. Although U.S. apparel manufacturers may not compete advantageously with the Far East in low-priced product categories due to the competitive prices and aggressive market strategies of the Far East, there are excellent opportunities to gain market share in the moderate to high end (quality-oriented) of the Mexican market.

Additionally, affluent Mexicans are concentrated around the cities of Mexico City, Guadalajara and Monterey, making this consumer sector easy for apparel manufacturers to target.

Much of Mexican market growth has occurred along the U.S. border, where maquiladoras have enriched local economies. Besides having a higher standard of living, the population of the northern region of Mexico is more 'Americanized' and more likely to be attuned to U.S. brands and advertisements than that in the less developed southern regions (Peterson, 1991). However, the pattern of prosperity is expanding southward as industries extend their operations to the south. The free trade agreement would benefit interior regions in much the same manner as the maquiladoras have in the north (Peterson, 1991).

President Salinas of Mexico began a recent presentation to The Business Roundtable as follows: 'Politically, the end of the cold war, the establishment of new alliances, the changes in Eastern Europe and the emergence of new democracies in Latin America and other regions of the world outline an agenda radically different from that of the past decade. In the economic sphere, new financial and technological innovation centers, the globalization of capital and commodity markets and the formation of new regional economic blocks are stepping up competition and heightening interdependence among nations and among enterprises' (June, 1990). Nowhere is the need for interdependence and cooperation more apparent than in the apparel industry.

19.3 NAFTA: U.S.–MEXICAN TRADE

NAFTA, a vital component in the U.S. national strategy to improve competitiveness in the international marketplace, will provide U.S. manufacturers with access to 85 million Mexican consumers (Galvez, 1991). U.S. manufacturers will have enhanced opportunities to export through direct exports, processing exports (807 or 807A) and joint ventures. From 1986 to 1990, U.S. exports to Mexico more than doubled, generating an estimated 264000 U.S. jobs (Dowd, of letters 1991). Three major economic analyses corroborate that the U.S. will benefit from NAFTA in terms of exports, output and employment (Dowd, 1991). For example 70 per cent of each dollar spent by Mexicans on imports is on U.S. goods and 15 per cent of all spending is on U.S.-produced goods and services (Dowd, 1991), in contrast with

Asia which spends almost nothing on U.S. goods. Mexico not only provides a lucrative market for U.S. exporters, it also provides an entrée into Latin America, which will become an increasingly important market as trade increases in the Western hemisphere.

NAFTA will create a North American market with a total output of $6 trillion, 25 per cent larger than the EC. Over 370 million people live in North America, compared with 320 million in the EC and 122 million in Japan. Thus opportunities for growth are considerable (Hughes, 1991). The free trade agreement could result in increased economic growth and jobs throughout the continent, lower prices due to improved efficiency and productivity and enhanced competitiveness of North American goods in the world marketplace as each nation brings its unique strengths into the partnership. The agreement offers an opportunity to create a grand economic zone that can match the economic power of Europe and the Pacific Basin.

A primary benefit of the agreement will be the increased potential to exploit the relative advantages of the three countries, their productivity and their ability to compete with Europe and Japan. For example, dismantling trade barriers will lead to expanded bilateral trade and more predictable rules for doing business across the border. Consumers in both countries will benefit from lower prices and a greater variety of products. Working together, North American capital, technology and labor could produce competitive products for the world market. This would allow U.S. textile and apparel manufacturers to increase their exports, not only to Mexico, but to the world's markets (Solana, 1991). Lower-cost Mexican labor averaged with higher U.S. wages for some of the value-added operations will allow for a more competitively priced product. Some manufacturers are already finding that integration of U.S. operations with production in Mexico is making their firms more internationally competitive.

There are close geographical, economic and social relationships between the United States and Mexico. One important way to capitalise on that relationship would be to develop a truly global and modern North American apparel manufacturing center.

19.4 GLOBAL COMPETITIVENESS: ASSESSMENT AND NEEDS

To become globally competitive, apparel businesses will have to continuously upgrade their manufacturing processes to increase the volume, quality, variety, customization and timeliness of 'product-to-market' delivery of competitively priced apparel. Japan and Germany, among other industrial nations, are basing their long-term economic plans on computers, high speed communications and universal education, a strategy which has resulted in new advances in productivity, quality, variety and speed of introduction of new products (Groves, 1990). This strategy is based on high-performance organizations which utilize the latest technologies and require (1) an educated and skilled workforce capable of implementing a participative and self-directed work culture and (2) global managers whose allegiance is to enhanced worldwide corporate performance.

Together the United States and Mexico have the natural resources, technology and essential human resources required to establish a globally competitive apparel industry. Much of the required infrastructure is already in place. Increased manufacturing in Mexico will lead to increased economic stability, thereby improving efforts to strengthen infrastructure investment in such areas as schools, roads and telecommunications. To be globally competitive, the new 'high performance' organizations will require a dependable and plentiful supply of labor. Given Mexico's youthful population and positive demographic profile, combined with the mature and productive workforce presently in the United States, the creation of a globally competitive apparel industry seems probable (Groves, 1990).

In order for the North American textile and apparel industry to achieve a globally competitive workforce, career ladders that extend from elementary schools through high school, technical schools and universities combined with on-the-job training and education are needed. The German model, where people aged 15 to 18 serve apprenticeships which combine work and school in an integrated effort to prepare students for an increasingly technological workforce, should be examined for possible application in North America. Certainly industry internships, sabbaticals to spend time in industry, and visits with educational and industry leaders worldwide would enhance the ability of educators in North American universities to provide their students with accurate and timely career information and shrink the distance between 'those who do and those who teach'. In addition, bringing industry leaders into the classroom and students into industry (via field trips, internships and cooperative education activities) would provide students with many of the competencies necessary for success in a globally competitive industry.

19.5 OPPORTUNITIES THROUGH NAFTA

If the North American apparel manufacturing industry is to thrive, or even survive, in the global market of the 1990s, changes will have to be made to meet the requirement for faster response, greater variety, consistently high quality and increased productivity. The U.S. apparel industry must deal effectively with the opportunities that have been opened up by recent global events by exploring international opportunities and positioning companies to take advantage of those opportunities as they occur. Information technology and the new realities of political economy are moving us into a time when integrated regions, not autonomous national markets, are the basic building blocks around which global companies organize (Sanderson and Hayes, 1990).

Internationalism presents opportunities as well as threats. Establishing wholly-owned subsidiaries, joint-venture operations or sub-contracting in low-cost countries are ways of exploiting geographical shifts. Cifra's deal with Wal-mart and the deal between Commercial Mexicana and Price Club are two noteworthy alliances which have been formed thus far. A number of U.S. apparel manufacturers and retailers are positioning themselves to capitalize on the market potential offered by Mexico.

As Mexico develops further, there will be an unprecedented growth in demand for many goods and services. U.S. companies are in an excellent position to supply this growing market, where consumers are eager to purchase U.S. goods. Presently the United States has the largest market share (54 per cent); however this share is slipping as Hong Kong and other Far East countries are aggressively targeting Mexico's apparel market. The United States future market share will depend on how aggressively U.S. apparel manufacturers address the needs of the consumer market in Mexico. Continued growth in Mexico will result in infrastructure upgrading and many new opportunities for U.S. textile and apparel companies to participate in modernization efforts. NAFTA will enable U.S. companies to take advantage of this largely untapped Mexican consumer market.

However U.S. exporters must keep in mind that the Mexican market differs markedly from that of the United States Purchasing power is smaller for most Mexicans, income groupings are different and personal contact is very important in launching a successful sales campaign. Additionally, potential exporters must be aware of the cultural differences that affect trade.

As in any international context, companies must adjust their marketing mix and business practices to be compatible with the norms, values and traditions of the market they are serving. Market growth depends on responding to local differences in the key markets around the world. Adapting to local differences requires a multifaceted organizational structure (Howard, 1991). Although some fashion products are global, many must be customized to meet the demands of the specific markets of destination. Through their maquiladoras and 807 operations and some retailing experience, many U.S. apparel companies have already made extensive advance in learning about Mexican culture and understanding the way things work in Mexico.

19.6 CONCLUSION

The apparel industry is divided over the proposed free trade agreement with Mexico with many fearing that eliminating quotas and tariffs with Mexico would disadvantage the U.S. apparel industry. Certainly, there will be some shifting of lower value-added jobs from the United States to Mexico, which will affect some workers in the labor-intensive apparel-manufacturing industry. In fact, the free trade agreement could further erode U.S. apparel production if the industry is not prepared to make the changes necessary to compete globally. The U.S. apparel industry must position itself for global competitiveness; not merely strive to maintain its current market position.

20 The Potential Effects of NAFTA on the Textile and Apparel Industry in the United States

Kathleen Rees, Jan M. Hathcote and Carl L. Dyer

20.1 INTRODUCTION

The prospect of the enactment of the North American Free Trade Agreement, which would result in the elimination of both tariff and nontariff barriers to trade between Canada, Mexico and the United States, has ignited concern within the textile and apparel industry. Historically this has been one of the most highly protected industrial sectors, and debates are taking place regarding the manner and extent that North American free trade will impact on the industry.

This chapter will present key issues relevant to the proposed trade agreement that are of major concern within the industry: (1) industry structure in the United States, (2) the textile and apparel industry in Mexico, (3) specific sectors and groups within the United States which will emerge as a 'winners' or 'losers' under the proposed agreement, and (4) some potential global impacts and ramifications of the agreement.

20.2 STRUCTURE OF THE U.S. TEXTILE AND APPAREL INDUSTRY

The United States textile, apparel and fiber industries together form one of the most complex and heavily regulated manufacturing sectors of the U.S. economy. The sectors are labor intensive, requiring relatively simple and inexpensive technology and a predominately semiskilled or unskilled labor supply, whose wages in the United States are about 70 per cent of the all-manufacturing average. The resulting low wages, low training costs and low capital costs help to provide comparative ease of entry into textile and apparel manufacturing industries, particularly apparel. These factors, combined with an excess supply of labor at wages significantly lower than in developed nations, are major reasons why most developing nations can, and probably should, move into manufacturing via textiles and apparel for domestic consumption and for export to enhance economic growth. Movement out of textiles and apparel into more sophisticated technologies are typical as growth occurs. Notable exceptions are the EC and the United States who, despite high-tech dominance, have maintained huge textile/apparel complexes through protection.

In the United States, this complex constitutes the single largest employment sector of the domestic manufacturing economy, employing almost two million blue and white collar workers in over 20 000 manufacturing facilities which collectively ship more than $100 billion in textile and apparel products annually. Of this total, about $7 billion were exports in 1991, in contrast with $34 billion of imports, leaving a trade deficit of over $26 billion on the textile/apparel complex. Double-digit deficits have existed for this complex since the 1960s.

Textile, fiber and apparel industry employment comprises approximately 10 per cent of total U.S. manufacturing employment. Loss of jobs to imports lies at the heart of the appeal by the textile/apparel complex for continued, and perhaps more comprehensive, protection even though textiles and apparel are among the most highly protected sectors of the U.S. economy. The textile/apparel complex has historically sought political solutions for what is essentially an economic problem for them. Therein lies much of the complexity of the textile and apparel component of NAFTA.

20.3   THE APPAREL INDUSTRY IN MEXICO

Apparel production was one of the early entrants into Mexico's maquiladora program. Designed to provide employment for Mexico's abundant labor force and an infusion of needed foreign capital into the country's economy, Mexico's in-bond program provided an ideal setting for the textile and apparel industry. Being highly labor intensive, yet requiring little capital outlay, apparel production facilities could readily be established across the border by U.S. manufacturers.

Traditionally, 30 per cent of all maquila manufacturing involved the textile and apparel industry. Currently approximately 10 per cent of maquila output consists of textile and apparel products, manufactured in approximately 275 plants involved in textile and apparel production in the maquiladoras. In 1988 over 200 plants were involved in apparel-assembly operations, employing 35 000 laborers constructing garments for U.S. companies (Black, 1991a). Since Mexico possesses little competitive advantage in textile production compared with the United States, much of the textile input for these plants has come from the United States. Low labor wages and proximity to the United States have therefore resulted in Mexico drawing apparel production from the Far East and becoming established as a 'Taiwan in the U.S.' backyard' (Craig and DuPont, 1989, p. 107).

The Mexican government's published average wage for apparel assembly workers is approximately $0.88 per hour, including fringe benefits (Black, 1991c). This contrasts with the higher wages in the United States as well as other countries, such as Korea, Taiwan and Hong Kong where a significant amount of apparel assembly/sourcing has traditionally occurred. The cost of transferring U.S. management to Mexico is much less than for alternative locations. Mexico's apparel production costs are lower than in the United States and it has been estimated that, taking the differences in labor and operating costs into account, U.S. firms can save approximately $15 000 per worker per year by locating in Mexico (Magnier, 1989).

Mexico's proximity to the United States provides another important advantage, allowing ease of distribution and transportation and improved logistics. Additional benefits exist regarding both the transfer of and access to managerial and technical staff and ease in providing technical assistance for production facilities. Collectively, these have enabled production firms to maintain accelerated turn-around times, increased quality control and decreased transportation costs. In particular, for apparel, with an abbreviated life cycle and a high value-to-weight ratio, such considerations are especially important and should continue to provide a competitive advantage for U.S. apparel production in Mexico.

Supportive governmental involvement on both sides of the border, and tariff and trade policies conducive to production sharing, have augmented the volume of trade between the two countries. Preferential treatment under GATT, the MFA, Items 807/807A and the Special Regime have liberalized restrictions on textile and apparel production and resulted in increased trade across the border. Thus NAFTA would further enhance bilateral textile and apparel trade between the United States and Mexico, but it is not yet clear what the structure and magnitude of that trade will ultimately be. That remains an empirical question not considered expressly in this chapter.

20.4 POTENTIAL GAINS AND LOSSES FOR THE INDUSTRY

Some sectors within the textile and apparel industry will gain from the enactment of NAFTA, while others will not fare as well. It is likely that U.S. apparel workers and producers, and some textile producers involved in domestic production, will be negatively affected by the agreement, while U.S. producers who shift production to Mexico, those domestic textile producers who are sources for Mexican apparel producers, U.S. fiber producers, retailers and consumers will benefit from the agreement.

20.4.1 Structural Transfer of Labor in Apparel Production

In the short run, domestic industry may suffer a structural transfer of labor resulting from lower wages in Mexico and preferential access of apparel imports to the United States under NAFTA. Given the labor intensity and lack of capital investment required for apparel assembly, the potential for employment transfer across the border is greater than for textile production, which requires a more highly skilled labor force and enhanced technology.

The industry fears the passage of NAFTA will accelerate the shift of downmarket production to Mexico. Developments regarding the MFA and the Uruguay Round of GATT will impact on the movement of labor as well. Continuation of the existing worldwide quota system under the MFA would further enhance Mexico's advantage within North America. In contrast, with the successful completion of the Uruguay Round of GATT and a phase-out of quotas, other competitive textile and apparel-producing nations would possess advantages similar to those which NAFTA would offer Mexico (Awanohara, 1991).

Concern has been expressed by labor and trade union officials that, with the passage of NAFTA, increased production Mexico would negatively affect employment in the U.S. domestic industry. Thus U.S. jobs could be lost due to the existing wage differential and the competitive advantages gained through engaging in production in Mexico.

According to the American Textile Manufacturing Institute (ATMI), Mexico is the sixth largest supplier of textiles and apparel to the United States. With Mexican imports having tripled since 1980, Carlos Moore, vice president of ATMI, asserted that increased production in Mexico has been a significant factor contributing to the loss of over 400 000 jobs during the same time period. Other industry officials contend that the figure may be as high as 420 000. Even though these claims appear to be excessive, Mr Moore further alleged that the only things prohibiting Mexico from further increasing its textile and apparel exports, and thereby further eroding U.S. employment in this sector, are the existing protective tariffs and quotas (Barrett, 1991).

While the U.S. textile and apparel industry has lobbied aggressively for protection from a potential increase in Mexican imports resulting from NAFTA, examination of Mexican apparel imports indicates that Mexico does not appear to pose a significant threat. In 1990 Mexican apparel comprised 2.9 per cent of total U.S. imports under the MFA. In the same year Mexico failed to fill its allowable quota of 170 million garments, exporting only 130 million to the United States (Wu, 1991). Of the total volume of Mexican apparel exports to the United States in 1990, approximately 90 per cent were produced as either 807 or Special Regime (807A). In dollar terms this was over $500 million in apparel, with approximately $150 million and $302 million in 807 and Special Regime imports, respectively. Only $60 million were actual imports of apparel originating in Mexico (Antoshak and Leitzell, 1991).

20.4.2 Protective Measures to Compensate for Potential Industry Losses

In order to alleviate the potential damage to domestic industry from the enactment of NAFTA, industry officials have recommended that either textiles and apparel be entirely omitted from the phase-out of import restrictions or that an extended phase-out period of up to fifteen years be enacted for a reduction in tariffs and quotas to allow sufficient time for the industry to adjust to Mexico's competitive advantage. A ten-year phase-out, as in the Canada–U.S. FTA, has been considered insufficient by many in the industry. Industry officials have urged implementation of the Trade Adjustment Assistance (TAA) program and the Economic Dislocation and Worker Adjustment Assistance Act (as part of the Job Training Partnership Act) for retraining displaced textile and apparel workers; however, in the past TAA has been inadequate relative to industry unemployment.

20.4.3 Potential Increases for U.S. Apparel Producers Operating in Mexico

Although shifting apparel production to Mexico is likely to result in a loss of domestic employment, a number of U.S. apparel producers view it as a means of establishing and maintaining competitiveness in the world marketplace. Carl H. Priestland, chief economist for the American Apparel Manufacturers Association (AAMA), forecasts that with the lifting of Mexican tariffs, U.S. apparel exports to Mexico, which currently stand at $800 million per year, would increase significantly. Mr Priestland predicts that U.S. exports will triple with tariff abolition (Wu, 1991).

Operating on a continental scale could increase the United States' ability to defend domestic jobs against European and Asian competitors. Future security for U.S. jobs may be provided, especially if strong rules of origin help to promote reimportation of assembled products made primarily of U.S. components. Some 807 producers assert that since garments assembled in Mexico are cut in the United States, U.S. jobs are supported by Mexican assembly workers (Barrett, 1991).

The Haggar Company's experience in Mexico exemplifies the potential for U.S. companies to benefit from moving their production facilities across the border. Having invested heavily in Mexican 807A production with a technologically advanced plant, Haggar presently produces two million men's trousers designed for the U.S. market, as well as lines designed specifically for the Mexican consumers. Further expansion and production for both the U.S. and Mexican markets are planned. Haggar has explained that its move to Mexico was driven not by the wage differential, but by incorporation of U.S. products in offshore assembly, the decreasing U.S. labor force, and by the declining attractiveness of assembly line work to U.S. laborers.

Typical of companies doing business in Mexico, Haggar has found that the present duties imposed on merchandise sold in Mexico have forced them to set higher than preferred prices, causing the price of their menswear to fall into the upper end of the market. The lifting of existing Mexican tariffs, presently 20 per cent, would enable Haggar and similar firms to produce for the larger, medium-range Mexican market at more competitive prices.

20.4.4 Potential Gains for Textile and Fiber Producers

U.S. textile and fiber producers stand to gain from the implementation of NAFTA since they are likely to supply the fibers and fabrics used in Mexican apparel production. Mexico is currently the third-largest buyer of U.S. textile products (with mill orders of $515 million in 1990) and is likely to continue to purchase textile products from the United States rather than from the Far East. According to Larry Mounger, president of the National Association of Textile and Apparel (NATA), 'The NAFTA is an opportunity for American companies that are currently sourcing their product in Asia to take a look at Mexico as an alternative. American textile companies may find great new markets for their textile goods in Mexico and Canada, thereby increasing our exports' (Black, 1991b p. 121). Recent trade missions sponsored by the U.S. Department of Commerce have established that increased opportunities for the expansion of the textile trade do exist, specifically with Mexico (Dawson, 1990, 1991).

The United States possesses a comparative advantage due to its sophisticated technology and automation. Thus U.S. production could be enhanced – rather than threatened – as increased production is undertaken by Mexico, augmenting domestic industries and job opportunities.

U.S. textile and apparel firms would gain from the reduction of trade barriers following the implementation of a free trade agreement. Experience with 807, 807A and the Special Regime has shown that apparel production in Mexico uses U.S. textiles, and the technical advantages possessed by U.S. textile producers leads to expectations that Mexican producers of apparel will increase imports of textiles from the United States as production increases. In 1990 the United States accrued a $16 million bilateral textile trade surplus with Mexico (McAllister, 1991), which should expand significantly with the advent of NAFTA.

To protect their interests, U.S. textile producers support strong rules of origin. Their concern is that East Asian manufacturers may consider Mexico to be a promising point of entry to the United States and thus engage in transhipping through Mexico. Therefore apprehension exists regarding the possibility of goods entering the country under the preferential treatment granted to Mexican goods through a free trade agreement. AAMA and ATMI have proposed that to qualify for dutyfree treatment under NAFTA, apparel must be produced within North America from yarn onward. Fiber manufacturers would prefer a fiber-onward rule, which would dictate that the natural or man-made fibers used should be of United States, Mexican or Canadian origin. For apparel producers, a fiber-onward rule of origin could be problematic, since certification of the origin of the fibers could prove difficult, if not virtually impossible, to verify.

20.4.5 Implications for Retail

Continued economic development of Mexico will augment the per capita income of its population and increase the nation's standard of living. As Mexico's inflation rate continues to decline, and as the value of the peso appreciates relative to the dollar, U.S. imports should become more price-competitive with Mexico's domestic products. Reduction of trade barriers will enable U.S. firms to export goods which are often less expensive and of better quality than those produced in Mexico.

As the Mexican economy continues to stabilize and expand, consumer spending will continue to grow. Opening Mexico's markets through free trade with the United States will allow access to 85 million Mexican consumers who are anxious to buy American-made products. This will be a diverse and rapidly growing market, estimated to reach 100 million by the turn of the century.

Mexican retail interest is increasing for medium- and upper-medium-quality apparel. Increased retail opportunities will be found primarily in urban areas. Since Mexican customers are conscious of both quality and brand-names and prefer American-made goods, opportunities exist for further expansion into the Mexican market with the enactment of a free trade agreement. With a reduction in bureaucracy and a lifting of the costs involved in exportation, U.S. goods can be offered to Mexican consumers at more reasonable prices.

A number of U.S. retailers, including Sears Roebuck, J.C.Penney's Units, Woolworth, and Wal-Mart (in a joint venture with CIFRA SA, Mexico's largest retailer), have already established operations in Mexico. Several companies are currently expanding their operations or planning future growth. Wal-Mart plans further expansion in Mexico by opening wholesale clubs similar to the Sam's Clubs it operates in the United States. Hartmarx Corporation intends to extend its operations in Mexico by using vertical marketing; the company will open a production plant from which it will ship to both its U.S. and Mexican stores (Barrett, 1991).

Representatives in retailing favor less cumbersome rules of origin and have proposed that all textiles and apparel be subjected to a single rule of origin, rather than sectoral handling. Accordingly they have proposed that goods be considered of North American origin if they are obtained or produced totally within North America or are subjected to a transformation that changes the tariff classification. When enforcing the rules of origin, such enforcement should neither impede unnecessarily the transfer of goods across the border nor require excessive paper work. Time delays and problems in receiving transborder shipments could result in the benefits accrued being lost at customs.

Those who are expected to benefit most under NAFTA, however, are consumers in Canada, the United States and Mexico. Consumers will gain, since freer trade will provide enhanced access to a greater quantity and variety of products and products that are lower in price.

20.5 CONCLUSIONS

Implementation of the North American Free Trade Agreement will create a significant counterbalance to the European unification of 1992. Combining the production of Canada, Mexico and the United States will create the world's largest trading bloc, with an anticipated GNP of $6 trillion, 25 per cent larger than that of the EC. With a combined market of approximately 360 million consumers, great promise exists for producers, suppliers and retailers within North America as well as for the economic health of each nation.

NAFTA will not be negotiated in isolation; it will affect domestic industries as well as industries in other countries. Trading partners, such as the Far East and the Caribbean, will be influenced by decisions made during the negotiation of the agreement. Proposed future trade policies, such as the Enterprise for the Americas, deserve careful consideration whilst NAFTA is being negotiated and implemented. The North American free trade agreement is a concept whose time has come.

Part V
Environmental Issues

21 Introduction

The primary purpose of this volume is to provide a balanced view of the prospects and problems of a North American free trade agreement. The preceding chapters have analyzed NAFTA from a historical perspective to labor issues, from national perspectives to bilateral issues, and from broad macroeconomic perspectives to bilateral issues. The perspective left for this final part is that of the environment. In Chapter 22 John J. Audley presents the environmentalist's view of NAFTA. First he examines recent environmental concerns over trade policy, and how those concerns have been translated into substantive objectives for trade negotiations. He then presents a critique of the 'green language' in NAFTA, based on the objectives identified by environmentalists. The chapter concludes with an examination of the negotiated outcome of NAFTA.

In his analysis Audley maintains that NAFTA 'reflects a departure in conventional trade policy because it attempts to address some of the key environmental concerns relative to trade'. However, he argues that inclusion of certain environmental clauses in NAFTA is far from sufficient and, in effect, 'environmentalists have failed to achieve the objectives they set out to achieve in trade negotiations'. To overcome this, Audley points out that the ultimate impact of NAFTA will be determined by Congress, and he proposes that the groups involved in trade should now refocus their attention to work with elected officials in the next stage of negotiations dealing with enabling legislation and other parallel agreements that accompany NAFTA.

22 The 'Greening' of Trade Agreements: Environmental 'Window Dressing' and NAFTA

John J. Audley*

22.1 INTRODUCTION

Historically the public debate surrounding trade policy has boiled down to a discussion about jobs. Trade advocates have argued that increased trade in U.S. goods creates jobs for U.S. workers. Organized labor has argued that trade agreements pit higher-paid U.S. workers against lower-paid workers from developing countries. The resulting trade policy has been based less upon available statistical evidence than on the ability of interest groups to influence the political process surrounding negotiations.

Now, trade agreements have another issue to consider. The recent debate surrounding the creation of the North American Free Trade Agreement (NAFTA) marks the first time that environmental issues have been addressed when forming U.S. trade policy. As organized representatives of environmental concerns, U.S. non-governmental environmental organizations (NGOs) possess adequate political resources to alter the focus of trade policy to integrate trade and environmental policy in NAFTA. Now that the negotiations have been completed† we can begin to evaluate that influence, and examine the factors which shaped the outcome.

* The author is currently trade analyst for the Sierra Club Center for Environmental Innovation. The arguments in this paper do not necessarily represent the views of the Sierra Club or its members. I am particularly thankful to Ruth Hennig and the John Merck foundation for financial support. All errors and omissions are the responsibility of the author.
† Formal negotiations between countries were completed 12 August 1992. The text of the agreement was submitted to Congress on 12 September 1992. The delivery of the text to Congress began the first phase of 'fast track'. Congress must now work with the Administration to develop the package of 'enabling legislation', attend to the changes to domestic law necessary to conform with the conditions in the Presidential Agreement, and present these changes as a bill to be voted on by both houses (see Destler, 1992).

NAFTA reflects a departure in conventional trade policy because it attempts to address some of the key environmental concerns relative to trade. Efforts to protect domestic environmental laws from preemption through harmonization, and language designed to mitigate the negative environmental impacts of expanded trade, are contained within NAFTA. Perhaps the greatest achievement was to increase public awareness of the interconnection between trade and the environment. But despite their new-found political leverage on trade policy, NGOs have not been able to reshape the focus of trade agreements from economic growth to 'sustainable development'. Environmentalists have failed to achieve the objectives they set out to achieve in trade negotiations.

This chapter will examine recent environmental concerns over trade policy, and how those concerns have been translated into substantive objectives for trade negotiations. It will then provide a critique of the 'green language' in NAFTA, based upon the objectives identified by environmentalists. Finally it will examine the negotiated outcome and explain some of the reasons why environmentalists did not achieve their goals in trade policy.

22.2 DEVELOPING THE ENVIRONMENTAL PERSPECTIVE ON TRADE: DEVELOPMENT vs GROWTH

Since the creation of an international trading regime under GATT in 1947, trade agreements have held unrestricted commerce as their ultimate goal. Early efforts to achieve this goal were focused on removing formal barriers to market access, such as tariffs on imported products. Beginning with the Tokyo Round in 1973, GATT members turned their attention toward 'non-tariff barriers', barriers such as technical requirements for products that are designed merely to protect domestic consumers from foreign competitions (Destler, 1992) by focusing on both tariff and non-tariff barriers, GATT trade rules work toward integrating the economies of the world into a single global market.

Trade philosophy acknowledges that restrictions will be necessary. These 'exceptions' to the principle of unrestricted trade must be justified, however. Article XXb of GATT allows restrictions to trade when it is 'necessary' to protect human, animal or plant health. Whatever the criteria used to determine what is 'necessary', GATT requires that all exceptions be 'least trade restrictive'.

A philosophy that requires trade restrictions to 'prove' themselves to be legitimate exceptions is problematic for environmental laws. Environmental laws and regulations are often designed to actually restrict trade, or they prohibit the consumption of products exposed to harmful products.* These laws are often not based upon the economic or scientific tests required by GATT, but instead reflect social values which are often inconsistent with 'scientific standards' (Sagoff, 1992).

To a great degree this apparent inconsistency between environmental and trade goals stems from the context in which GATT was first written. GATT originated in the 1940s, a time when the world was unaware of the environmental implications of trade. As a result, nowhere in the original GATT agreement is the word 'environment' used. But regardless of its origin, the net effect of GATT's ignorance of the environmental implication of trade has been to apply the same 'exceptional' principle used to remove trade barriers to domestic environmental regulatory regimes (Wirth, 1992).

Little has been written about these concerns until recently. Although general concern regarding the impact of global business on the environment extends back to the 1960s, formal awareness of the interconnection between trade and the environment was first identified in the 1987 Brundtland Commission Report on Environment and Development. The Commission questioned the relationship between international trade and environmental protection, calling for reform of the international trading system under GATT (WCED, 1987). The Uruguay Round of GATT has failed to heed this message however, as environmental issues are not part of its mandate.†

As part of its report, the Commission argued that trade should be based upon 'sustainable development'. International commerce should be promoted with the goal of 'meet[ing] the needs and aspirations of the present without compromising the ability of future generations to meet their own needs' (WCED, 1987, p. 43). Although still a developing philosophy, two principles have arisen. First, sustainable development distinguishes between growth and development. Goodland, Daly and El Serafy see the distinction in terms of the scale and

---

* Examples include the Endangered Species Act and the Delany Clause in the Federal Food, Drug and Cosmetic Act.
† In its defense, the mandate for the Uruguay Round was finalized in 1986, before the Commission issued its report. Changing the mandate for negotiations is not impossible however. Intellectual property, not originally a part of the 1986 mandate, is one of the most important aspects of Uruguay Round negotiations.

quality of human activity: 'When something grows it gets quantitatively bigger; when it develops, it gets qualitatively better' (1991, p. 85). Second, sustainable development promotes greater public participation in the policy process. As policy decisions are increasingly made by technical administrators, who are difficult to hold accountable for their actions, greater procedural transparency is essential for public acceptance of their regulations, and to 'watch the watchers themselves' (Hardin, 1968, p. 1246).

Quality and scale of human activity and increases in procedural transparency were the underlying goals of NGOs as they engaged the Bush administration in discussions surrounding NAFTA. The task facing them was to transform these rather broad objectives into substantive policy recommendations.

22.3 ENVIRONMENTAL MESSAGE AND U.S. TRADE POLICY

Until recently U.S. national environmental organizations were ill-prepared to enter trade discussions. The relatively new interest in exploring the interconnection between trade and the environment, and the lack of technical expertise, made it difficult for NGOs to focus both financial and human resources on trade (Wathen, 1992, p. 71). Beyond the usual problems associated with developing new policy campaigns, there were numerous institutional barriers to overcome. Trade negotiators did not provide an avenue for environmental participation as trade policy had never before been criticized by environmentalists.

An avenue was created, however, when discussions of expanded trade between Mexico and the United States focused attention on the environmental problems on the U.S.–Mexico border region. In a report by the American Medical Association, the border region was described as a 'virtual cesspool and breeding ground for infectious diseases' (*JAMA*, 1990, p. 3320). Much of the blame was placed on the rapid industrialization resulting from the Border Industrialization Program (BIP), which lacked accompanying increases in urban and social infrastructures. Both Presidents Salinas de Gortari and Bush seemed sensitive to the connection between expanded commerce between Mexico and the United States and environmental degradation. President Bush's request for fast-track negotiating powers coincided with the announcement of binational efforts to development a new agreement to address common-border environmental concerns.

A conference held in Washington DC in December 1989, which was cosponsored by the Mobilization on Development, Trade, Labor and the Environment (MODTLE) and the National Wildlife Federation (NWF), marked the first time that interest groups had focused congressional attention upon the integration of trade and environmental issues. At that time the environmental focus on trade policy was defensive; NGOs sought protection from the negative impact of trade: 'no additional harm to the environment from expanded trade' (Ward, September 1992a). More recently the focus has shifted toward ensuring that trade agreements actually promote environmental protection (Sierra Club, 1992). However, this more defensive approach, 'no more harm to the environment', shaped the introduction of environmental issues to NAFTA trade negotiations.

Institutional barriers combined with environmental inexperience in trade policy retarded NGO participation in NAFTA. Three subsequent events triggered greater attention; the fight over fast-track authorization, the release of the 'Dunkel Final Act Text' of GATT, and the Tuna/Dolphin GATT panel decision.

'Fast track' is the term used to describe elaborate legislative procedures created by Congress to facilitate the passage of trade agreements (Destler, 1992). The president requested fast-track authorization for NAFTA in January 1991. Concerned by the potential harmful effects of trade with Mexico to jobs and the environment, Senator Lloyd Bentsen and Congressman Dan Rostenkowski wrote to the president on 3 March 1991 asking him to address these issues. The president's May 1991 'Response to Congressional Concerns Arising from the North American Free Trade Agreement' did not satisfy the concerns of most NGOs following fast track. Although publicly supported by the NWF, the request for fast track was opposed by 16 environmental and consumer organizations (Wathen, 1992, see 'directory').

The second event involved the release of the December 1991 'Final Act' Text of the Uruguay Round of GATT. Analysis provided by Public Citizen detailed the environmental implications of the Uraguay Round (Wallach, 1991). Environmental concerns were further inflamed when GATT director General Arthur Dunkel published his own *Report on Trade and the Environment*. Dunkel argued that trade was not responsible for environmental problems, it just 'magnified' the existing inabilities of domestic and international environmental policies to achieve their goals. Dunkel expressed concern over environmental efforts to restrict trade because he felt it actually harmed their own efforts. Expanded trade increased the level of available resources to address problems of the environment (Dunkel, 1992, p. 2).

But the event which sparked greatest alarm involved international attacks on existing U.S. legislation. In the fall of 1991 a GATT dispute panel ruled that domestic efforts to protect the global commons are not permissible. The now famous 'Tuna/Dolphin' decision between the United States and Mexico (GATT Panel Report, 1991) expanded environmentalist's leverage on trade policy. Taking advantage of the leverage created by the decision NGOs were able to convince the House of Representatives to pass a non-binding resolution expressing their concern to the president. House Concurrent Resolution 246, the Waxman/Gephardt Resolution, urged the president not to present Congress with any trade agreement which harmed existing environmental, health, safety and labor laws. Begun in the fall of 1991, Waxman/Gephardt passed the House by unanimous vote (308 to 0) in July 1992.

By March 1992 the principle environmental NGOs began to develop concrete proposals for negotiators. Their positions were largely in response to the agenda defined by negotiators. Their cues came from a number of sources. Media attention was focused on NAFTA, and routine articles appeared in the London *Financial Times* and the *Journal of Commerce*. Additional information was provided by NGOs acting as private advisors,* and through public meetings with negotiators from the Office of Trade Representative (USTR). In addition, a leaked version of the agreement, dated February 1992, enabled NGOs to analyze the text and publicly criticize its tone and direction (Audley, 1992b).

NGOs produced three documents which detail their concerns: the NWF/Pollution Probe-Canada 'Bi-national Statement of Environmental Safeguards that Should be Included in the NAFTA' (May 1992); the Natural Resources Defense Council (NRDC) 'Environmental Safeguards for the North American Free Trade Agreement' (June 1992); and the Center for International Environmental Law (CIEL) 'Preliminary Overview of Environmental Concerns Arising From the NAFTA' (August 1992). More than twenty groups assisted in the drafting of these documents. Together the documents encompass the full range of environmental concerns in trade policy.

---

* These included Jay Hair, the National Wildlife Federation, John Adams, the Natural Resources Defense Council, Peter Bereley, the National Audubon Society, Russell Train, the World Wildlife Fund and John Sawhill of Nature Conservancy.

Environmental concerns can be categorized into three broad areas:

1. Protection against preemption of domestic laws by international agreements.
2. Procedural transparency in the administration and dispute resolution of the agreement.
3. Enforcement of environmental regulations.

A fourth area of concern, clean-up of the U.S.–Mexico border region, takes advantage of administrative interest in trade to address long-overdue environmental problems.

To implement their goals, NGOs recommended that NAFTA should create a trinational environmental commission to act as an investigative body capable of monitoring and reporting on regulatory enforcement and compliance. In addition, the Commission would assist citizens seeking financial relief from the failure to enforce existing environmental law. To aid enforcement, NGOs also called on NAFTA to generate financial resources dedicated to mitigating the negative effects of trade.

These conditions – procedural transparency, protection against preemption, enforcement, border clean-up and adequate funding – became the foundation of NGOs' substantive proposals to negotiators. How NAFTA stacks up to these conditions can act as a measure of environmental influence in trade policy.

22.4   ENVIRONMENTAL QUALITIES OF NAFTA: GREEN TRADE OR ENVIRONMENTAL WINDOW DRESSING?

The 12 August 1992 Presidential announcement of the completion of negotiations over NAFTA included a detailed defense of the 'greenness' of the agreement: 'NAFTA also is the first trade agreement to include far-reaching provisions to protect and improve the environment' (White House press release, 12 August 1992). Administrative officials outlined ten significant environmental provisions in NAFTA:

1. Commitment to sustainable development (Preamble).
2. Protection of environmental standards (chapter 7: agriculture; subchapter B: sanitary and phytosanitary (S&P); chapter 9: standards-related measures (TBT)).
3. Ability to maintain higher than international standards (S&P subchapter).

4. Promotion of upward convergence of S&P regulation.
5. Encourages enforcement of standards (Articles 759, 764).
6. Protection of international environmental agreements (Articles 103 and 104).
7. Enhances environmental protection through investment (Article 1114).
8. Provide access to environmental experts in dispute resolution (Articles 2014, 2015).
9. Facilitated land transportation will reduce air pollution along border.
10. Access to Mexican markets for U.S. environmental engineering companies.*

An important addition to this list occurred on 30 September 1992. EPA director Reilly announced the Administration's intention to create a trinational environmental commission, charged with addressing the environmental concerns related to trade.

If NAFTA were to achieve the goals expressed in its 'green language' it would mark a revolution in the nature of trade agreements, and the linking of trade policy with concerns for regional environmental protection. But when compared with the objectives laid out in the three NGO documents, NAFTA is failing to meet the environmental condition of 'no more harm'. Clearly the NGOs were unsuccessful in achieving their specific agenda for trade policy, yet the president still felt able to call NAFTA 'green'. The following is a more detailed analysis of the text.

NAFTA represents a departure from conventional trade agreements in that it attempts to address concerns regarding the preemption of domestic environmental laws. It attempts this in two ways. First, in the Preamble to the text, NAFTA makes a commitment to sustainable development. Second, NAFTA stipulates that parties have the right to set their own regulatory standards:

> in accordance with this Subchapter, adopt, maintain or apply any sanitary or phytosanitary measure necessary for the protection of human, animal or plant life or health in its territory, including a measure more stringent than an international standard, guideline or recommendation (Article 754(1)).

---

* Summary based upon William K. Reilly, Administrator to U.S. Environmental Protection Agency testimony before Senate Finance Committee, Subcommittee on International Trade, 16 September 1992.

Each Party may, in accordance with this Agreement, adopt, maintain and apply standards-related measures, including those relating to safety, the protection of human, animal and plant life and health, the environment, and consumers, and measures to ensure their enforcement of implementation. Such measures include those to prohibit the importation of a good of another Party or the provision of a service by a service provider of another Party that fails to comply with the applicable requirements of such measures or to complete its approval procedures (Article 904(1)).

On the surface it appears that environmental concerns are addressed in these two provisions. However, examination of the 'necessary' clause, and of what it means to be 'in accordance with this Agreement' reveals serious loopholes in the language of the text.

As discussed earlier, the 'necessary' clause originates in GATT Article XX(b). The application of this test has consistently overturned regulations which block market access.* According to one legal scholar, 'I am unaware of a single GATT panel report that turns on the interpretation of this word where the measure in question was held to be consistent with GATT' (Wirth, 1992, p. 10). To be 'in accordance' with the agreement, the level of environmental protection must be an 'appropriate' level (Article 754(2), (5)). Appropriate levels are defined in Article 757, 'Risk Assessment and Appropriate Level of Protection'. Part of the risk assessment criteria includes a requirement that a regulation be based upon 'scientific evidence' (Article 754(1b)). But as in the case of the Delaney Clause, this test is difficult for some environmental regulations to pass.

NAFTA also attempts to protect existing international environmental agreements (IEAs) from actions taken under the terms of the agreement. Article 104 identifies three IEAs† which will prevail in the event of any inconsistencies between NAFTA and the terms of the accords. However, a further clause in the text places these IEAs at risk. The terms of the IEAs shall prevail,

---

* Including the 'Malted Beverage Case', and the 'Thailand Cigarette Case'. For details on these cases, and complete GATT dispute panel references, see Charnovitz, 1992, and Christensen and Geffin, 1992.
† Convention on the International Trade in Endangered Species of Wild Fauna and Flora, the Montreal Protocol on Substances that Deplete the Ozone Layer, and the Basel Convention on the Control of Transboundary Movements of Hazardous Wastes and their Disposal.

provided that where a Party has a choice among equally effective and reasonable available means of complying with such obligations, the Party chooses the alternative that is the *least inconsistent* with the other provisions of this Agreement (Article 104(1d)) (emphasis added).

The phrase 'least inconsistent' is an attempt to overcome problems with a principle in GATT called 'least trade restrictive'.* This principle requires that regulations prove that there are no other, less trade-damaging, alternatives when achieving the environmental goal (Audley, 1992b). As a result of the severity of this test, 'least trade restrictive' is a subject of serious concern to NGOs. 'Least inconsistent' tries to avoid this problem by replacing the old principle with one created within NAFTA. However, 'least inconsistent' has no legal history, and will rely upon the panels created by NAFTA to determine its definition.

Because the interpretations of the language in the text are subject to disagreement, tremendous emphasis is placed upon the mechanisms created to resolve disputes. Here NGOs sought 'procedural transparency', where citizens would have the right to submit 'amicus' briefs, or be granted access to the proceedings (or the documentation). NAFTA attempts to address these concerns, not through greater transparency, but by guaranteeing that NAFTA, not GATT, will resolve environmental disputes. While a complaining party in NAFTA may select between GATT or NAFTA dispute proceedings, Article 2008(4) allows the responding party to insist on NAFTA *if* the dispute involves environmental issues. The provision allowing responding parties to demand NAFTA proceedings is an important improvement, given the record of GATT panel decisions regarding environmental issues.

Beyond this important exception, dispute proceedings are remarkably similar to those of GATT. NAFTA specifically allows panelists to seek out environmental experts or create scientific review boards when considering environmental issues (Articles 2014–15). These provisions are merely a restatement of a right that already exists, as panelists may call on whomever they wish for information or evidence.† In addition,

---

* Public meetings held between EPA officials and environmental organizations, 22 August 1992 (personal notes).
† Based upon notes taken during a meeting of the EPA Working Group on Trade and the Environment, spring, 1992. Members of the Working Group have served as GATT panelist. During the course of conversation the Group acknowledged that panelists' have the right to secure information from any source they wish. What is not possible is for non-contracting parties to submit documents without permission of the panel.

any additional information to the proceeding requires agreement from all disputing parties (Articles 2014–16). NAFTA does not improve procedural transparency. All aspects of the proceedings are confidential (Article 2017(3)). There are also provisions which will allow parties to block the publication of panel reports: 'Unless the Commission decides otherwise, the final report of the panel shall be published 15 days after it is transmitted to the Commission' (Article 2017(4)).

Enforcement of environmental laws was one of the most difficult issues for negotiators to address. Transboundary enforcement most notably raises the issue of national sovereignty, as environmental regulatory agents may require access to territory outside their national boundary. NAFTA attempts to address this issue through investment provisions in the agreement (Article 1114(1), (2)), and through side or parallel agreements that address the border (the Integrated Border Environmental Plan (IBEP)), and the recently proposed trinational commission on the environment.

The two parallel agreements are important attempts to address environmental concerns. IBEP is the product of President Salinas' and President Bush's January 1991 call for increased binational environmental efforts. It replaces an earlier agreement, the 1983 la Paz Agreement signed by Presidents Reagan and de la Madrid. But while taking important steps toward cooperative enforcement, greater citizen participation, and voluntary reduction, IBEP fails to overcome the historical problems plaguing border environmental issues (Rich, 1992). President Bush's request for $240 million for the first phase of IBEP was adjusted by more than 40 per cent by Congress, who argued that the funds were not available (Gephardt, 1992). In addition, Mexicans have failed to publish their version of IBEP, nor have they included Mexican citizens in the private advisory process.

The proposed trinational commission on the environment is still under discussion. Theoretically it would be headed by officials of all three national environmental agencies, whose purpose would be to provide a forum for discussion, a means for encouraging and supporting cooperation, and a mechanism for coordinating expertise and information (NWF press release, 30 September 1992).

The language in the investment chapter provides little environmental protection. Article 1114(1) guarantees the national right to maintain or enforce environmental measures appropriate to protect the environment. But in the event that parties fail to implement environmental regulations which might increase production costs, the investment chapter provides no avenue for dispute resolution. Article 1114(2) encourages parties to recognize that:

> it is inappropriate to encourage investment by relaxing domestic health, safety, or environmental measures . . . If a Party considers that another Party has offered such an encouragement, it may request consultations with the other Party and the two Parties shall consult with a view to avoiding any such encouragement.

Nothing in either of these two paragraphs is legally binding, nor does the language allow countries to seek damage relief through dispute process, as exists in the case of intellectual property (Chapter 17) or countervailing duties (Chapter 19).

Based upon the objectives laid out by NGOs, NAFTA fails to meet the 'no more harm to the environment' test. Table 22.1 compares NAFTA's provisions with the full range of conditions laid out by NGOs. Yet despite the performance of NAFTA relative to these objectives, to date the agreement has been endorsed by the NWF, and endorsements are expected from the World Wildlife Fund (WWF).* These endorsements are beginning to polarize NGOs, and ultimately weaken the leverage their issues have on trade policy.

22.5 ANALYSIS

A number of circumstances surrounding NAFTA shaped the degree to which NGOs were able to use their political leverage to influence NAFTA negotiations: public acceptance of the overall benefits of 'free' trade; administrative control over aspects of the political agenda; and the internal divisions among leading environmental groups on trade policy, which weakened their bargaining effectiveness. Less governable factors helped amplify their leverage. However, the final outcome of negotiations will not be determined until Congress is asked to vote on the enabling legislation accompanying NAFTA.

Americans have long been believers in the benefits of global commerce. As consumers of the largest portion of the world's goods,

---

* These statements, based upon examination of the testimony by Kathryn Fuller of WWF support my prediction. Ms Fuller states 'while WWF can endorse the NAFTA as a positive first step, we feel it essential that governments take a broader view of trade and environment' (Fuller, 1992, p. 2).

Table 22.1 Comparison of NAFTA with environmental objectives

	NRDC	CIEL	PP/NWF	NAFTA
Funds for trade-related abatement	Yes	Yes	Yes	No
Enforcement must be binding to avoid trade distortions	Yes Strict enforcement is needed	Yes Standards are useless without enforcement	Yes An enforcement mechanism must be in NAFTA	No NAFTA does not address this issue
Dispute panels open to public	Yes Public access is needed	Yes Input and access needed	Yes Input and access needed	No Tight secrecy prevails
Dispute panels open to outside experts	Yes Impartial experts should be involved	Yes Need to equalize trade and environmental concerns	Yes Include experts	Yes But only at the consent of all parties
Domestic standards protected from challenge	Yes NAFTA must protect domestic standards	Yes NAFTA must protect national and local standards	Yes NAFTA should protect highest possible standards	No U.S. national and local laws are open to challenge
Other IEAs have precedence	Other IEAs should be recognized	—	Yes Must protect rights to implement other IEAs	Maybe NAFTA language is unclear
Environmental impact statement (EIS) needed	—	—	Yes EIS should be completed on NAFTA	No Administration plans no EIS
Commission on the environment	Yes A trinational body is needed	Yes Trinational mechanism for planning and monitoring	Yes Need commission to investigate and monitor the environment	Maybe A body is being planned, but it was not in the Sept. 6 text

the integration of economies to take advantage of competitive advantages has pushed prices down and enabled us to consume more. So long as the negative aspects of trade have not fallen directly upon a single voter, or those problems have been mitigated through compensation, trade agreements have served to benefit American consumers rather than hurt them.

This inclination toward NAFTA was aided by a number of econometric analyses which predicted that jobs in the United States would increase as a result of NAFTA.* In an economic recession the prospect of greater employment is difficult for congressional leaders to oppose, even if their immediate constituents may be negatively affected by NAFTA.

The general belief in expanded trade is reflected in Congress's tendency toward general acceptance of new trade agreements. Destler argues that Congress endorses agreements because the constituents who are most responsible for their reelection (those which Destler calls 'elites') are benefited by expanded trade. Yet elected officials require more than the support of elites. Therefore, as they endorsed expanded trade, they also sought personal protection from the negative backlash associated with job loss. Their goal in creating special legislative procedures such as fast track was to 'make the buck stop somewhere else' (Destler, 1992, p. 15). Fast track shifted primary responsibility for the consequences of trade onto the president yet retained the ability to respond to interest group pressures to alleviate its more harmful effects through enabling legislation.

To further convince citizens of the benefits of trade, a new language was developed to internalize the positive aspects of trade. Trade agreements promote 'free' trade, a word that implies no costs to anyone. As more trade is 'costless', 'free' trade has become a mantra for American society. Anything that smacks of 'unfree' trade is bad. To be against 'free' trade is to be 'protectionist', a term which within the environmental community has very positive connotations. In its more current usage, the word 'protectionist' is something that is unthinkable in contemporary society. Mainstream elected officials consider it unacceptable to be opposed to the principles of 'free' trade.

President Bush was able to use the symbolism around trade and his administration's ability to influence the political agenda to mitigate

---

* A White House announcement referred to economic predictions of more than one million jobs being created by NAFTA by 1995.

environmental concerns for trade policy. The administration used the confidentiality surrounding the trade negotiations to keep opponents of his policy in the dark about time schedules or progress. Control over information enabled the administration to remain publicly opposed to linking trade and environmental policy until negotiations were complete. Announcing a 'green NAFTA' after refusing to integrate the two issues helped the president to minimize environmental demands while publicly appearing to embrace the environmental mandate.

President Bush took every opportunity to set the stage for a positive presentation of NAFTA. When NAFTA was announced, the *Washington Post, New York Times*, and *The Wall Street Journal* all ran editorials the following day in support of NAFTA. Attempts by groups like Public Citizen and the Sierra Club to quickly counter those endorsements were unsuccessful as these papers chose not to run the groups' editorial efforts. For national papers to run an editorial piece the following day it requires advance notice of the announcement, an advance notice not provided to opponents of President Bush's trade policy.

The president also used agenda control to include some environmentalists in private negotiations. Inclusion coopted their concerns with little cost to his overall objective. Environmental advisors numbered only five out of a total 1000. They were not permitted to discuss publicly the terms of the negotiations. By selecting organizations more supportive of his overall trade agenda, President Bush was accurately able to claim that NGOs had participated in negotiations, while at the same time muffling their public voices and excluding environmentalists with more fundamental concerns about trade.

Extraordinary circumstances like the Tuna/Dolphin decision and the poor performance by GATT with regard to environmental issues increased environmental leverage on trade policy. Symbolically Tuna/Dolphin revealed problems of national sovereignty; and the greatest environmental impact is in the area of preemption. Despite the concerns over the 'necessary' test or what determines 'appropriate levels of risk', NAFTA attempts to address these concerns. Removal of these troublesome phrases would help clarify negotiators' intentions and remove the lingering suspicions of NAFTA in these areas.

However, these adjustments in the president's agenda for trade were minor in comparison with his success in avoiding issues of procedural transparency and enforcement. Procedural transparency was not addressed. Enforcement through the creation of a trinational commission and the IBEP is, at this time, still just a goal. But with the support of organizations like the NWF, the administration is able to claim that President Bush kept his 1 May promise to Congress and produced the 'greenest trade agreement ever' (King, 1992).

But perhaps the biggest reason environmentalists were unsuccessful in pushing for more comprehensive change in NAFTA was their inability to speak in a single voice on trade. While the conditions spelled out in the three NGO documents were principled objectives for some, they were only negotiating points for others. Based upon Table 22.1, less than half of NWF's objectives were achieved, yet they were the first to endorse NAFTA.

Environmental organizations historically operate in loose coalitions surrounding policy issues. Efforts to build coalitions usually involve the creation of 'consensus positions', such as those generated during the NAFTA negotiations. Then it becomes a question of the degree to which organizations will adhere to the principles spelled out in these documents and remain united to support or oppose legislation. As each NGO has its own constituency, and each has little real influence over the behavior of the others, they lack the necessary incentives to induce or coerce unity.

NGO consensus was weak throughout the NAFTA negotiations. Most of the active groups in trade, the NWF, the WWF, and the Environmental Defense Fund (EDF) were 'conditionally supportive' of NAFTA, so long as some attention was directed at environmental concerns. They were also the groups involved in the private advisory process. Their more incremental or insider approach stands in contrast to organizations such as the Sierra Club, Friends of the Earth and Greenpeace, who began negotiations from a position of 'conditional opposition'; that is, opposition to NAFTA *until* the Administration offered sufficient environmental protection to win their support.

These two contrasting positions made the environmental community ripe for coalitional breakdown. Little or no cost was imposed on those groups who used the consensus documents as negotiating chips rather than principled positions. And, as insiders to the process, they were possibly able to gain more personal rewards by accepting the president's agreement than they would have done by achieving the environmental goal of sustainable development. If these benefits exist, they are not yet visible, or have not yet been conferred.

While NGOs have failed to achieve their philosophical goal of sustainable development, it is important to recognize that Congress, not the president, will ultimately determine the impact of NAFTA. The enabling legislation that must be written, any parallel agreements that accompany NAFTA, or a revocation of fast track privileges are roles for Congress. NGOs involved in trade will continue to work with elected officials in the next stage of negotiations. Given President Clinton's position on NAFTA (Clinton, 1992), it is certain that NAFTA as it exists, will continue to shape the final outcome. If this is the case when Congress reconvenes in January 1993, the goal of sustainable development in trade policy will remain a distant dream for environmentalists.

Part VI
Epilogue

23 Epilogue

Khosrow Fatemi

Much seems to have changed in the few weeks since the preceding chapters were written. In the United States, President Bill Clinton is now in office and in charge of the country. Both in the United States and elsewhere in the world, government leaders, business executives, and the masses alike are gradually getting used to the new style, if not yet the substance, of his administration.

In the international trade arena, the performance and pronouncements of the Clinton Administration during its first few weeks have at best been haphazard and confusing. Some of this was to be expected and attributable to the inexperience of the people in charge. Nevertheless, if the Clinton Administration's early performance – both rhetorical and substantive – is any indication, the future of free trade, as the principal theme of the American foreign economic policy, seems rather bleak.*

More significantly, the pivotal role of free trade as the dominant philosophy of the international trade arena, unchallenged during the past several decades, now seems to be on the decline – not only in the United States, but also in Europe and the rest of the world. In short, 'fair trade' seems to be replacing 'free trade', as the principle guiding light of international trade policy development around the world. The inherent danger of this new *modus operandi* is that it conceals the fact that for many, fair trade is a euphemism for protectionism.

In the case of the United States, the evolutionary rise of protectionism received a major boost from the electoral defeat of George Bush, a *bona fide* free trader. Consequently, in the case of the United States the timing of the change can be more readily identified.

---

* One of the very first actions taken by the new U.S. Trade Representative (USTR) after his confirmation by the U.S. Senate was to impose trade sanctions again the European Community. On February 1, 1993, Mickey Kantor, the new USTR, 'complaining that the EC had recently adapted Buy-Europe procurement rules, said it would block European companies from obtaining contracts from the federal government for telecommunications and power-generation equipment and a wide variety of services, beginning March 22 [1993].' 'U.S. Threatens Trade Action Against the EC,' *The Wall Street Journal*, February 2, 1993, page A2.

Furthermore, because the United States is the world's largest trading nation, the impact of the change can be more readily measured. The scope of the new protectionism, however, is by no means limited to the United States; nor is it limited to regional trade negotiations such as the North American Free Trade Agreement.

What is particularly threatening to the future of the free trade, more as a philosophy than as a policy, is the pervasiveness of the protectionism around the world: a fact, now amply evident in the United States. Even a cursory review of the popular press during the first few weeks of 1993 illustrates the accelerating nature of this problem. For example, according to a new analysis by *The Wall Street Journal*, 'As the Clinton Administration has shown in the past two weeks, governments get drawn into taking steps that restrict trade, usually to punish the "unfair" trade practices of somebody else. It's all done in the name of free trade, and always with regret. But each move further restricts yet another piece of international commerce and invites reprisals.'*

IMPACT ON NAFTA

So far as the North American Free Trade Agreement is concerned, President Clinton's position, while generally supportive, has been vacillating. On some occasions, such as his press conference following his meeting with President Salinas de Gortari of Mexico, he has been an ardent supporter.† At other times, he has expressed reservation about some aspects of the treaty as negotiated. The solution to addressing his concerns in the areas of labor laws and environmental issues, apparently acceptable to both Mexico and Canada, is to leave the NAFTA intact and negotiate the changes that the Clinton Administration wants in a series of concurrent accompanying agreements. Under this scenario, NAFTA, its implementing legislature, and

---

* 'Clinton Moves in Name of Free Trade may be Inviting Deterioration Instead', *The Wall Street Journal*, 5 February 1993, p. A8.
† In a press conference which followed his 7 January 1993 meeting with President Salinas de Gortari, Mr Clinton reaffirmed his support for the North American Free Trade Agreement by declaring his support for NAFTA and stating that there was no need to reopen the negotiations for the text of the Agreement. In fact he declared to speed up the ratification process, shortly after his inauguration, he 'would put one person in charge of organizing these issues'. 'Clinton Reaffirms Support for NAFTA', *Dow Jones News Service*, 8 January 1993.

all these side-agreements will be submitted to the U.S. Congress at the same time.

One logistical problem with this approach is the fact that the fast track authority which gave the Bush Administration the authorization to negotiate the North American Free Trade Agreement expires on 1 June 1993.* From the beginning, it was almost certain that the Clinton Administration would seek an extension of the fast track authority to complete NAFTA. In fact, on 11 February 1993, Mickey Kantor, the new U.S. Trade Representative announced that the Administration would indeed seek an extension of the fast track authority.† The extended time can also be used constructively to revive and seriously negotiate the completion of the Uruguay Round of the GATT (General Agreement on Tariffs and Trade) negotiations.

In conclusion, the North American Free Trade Agreement's ratification by the U.S. Congress, despite some strengthening opposition in recent months, still seems more likely than not. NAFTA's ratification and implementation would be a major step in slowing down the acceleration of the recent trend toward protectionism. It will not, however, be sufficient by itself to stop this trend, much less reverse it.

The resurgence of protectionism under the guise of 'fair trade' is a populist move with great appeal to the economically under-privileged groups such as the unemployed and those whose economic survival is threatened by what they perceive as unfair foreign competition. This trend could easily lead to one-up-manship in international trade and result in chaos and a repetition of the depression of the 1930s. To reverse it, that is to argue against short term benefits and populist political rhetoric would require political backbone and foresight among the world's leaders. Regrettably, that seems to be in short supply.

---

* From the negotiators' perspective, the primary advantage of the fast track authority is that the congress will not be able to make any amendments to the negotiated agreement. Each Congressman or Senator can vote for or against the agreement, but they cannot alter it by making amendments to it, attaching riders to it, or alter it in any form or shape. If NAFTA is not ratified before the fast track authority expires or is extended, the Congress is certain to want to make numerous changes in the agreement.
† 'Mr. Kantor said the Clinton administration will seek an extension of this so-called fast-track authority, under which Congress can vote "yes" or "no" on trade deals but not try to amend them. However, he didn't indicate when the administration will seek a renewal or for how long.' U.S., EC Agree to Press Japan on Trade Issues', *The Wall Street Journal*, 12 February 1993, p. A2.

Bibliography

ABOWD, JOHN and RICHARD FREEMAN, *Immigration, Trade and the Labor Market* (University of Chicago Press, 1991.

ADLER, M., 'Specialization in the European Coal and Steel Community', *Journal of Common Market Studies*, vol. 8, 1970, pp. 175–191.

AGMON, T., 'Direct Investment and Intra-Industry Trade: Substitutes or Complements?', in H. Giersch (ed.), *On the Economics of Intra-Industry Trade* (Tübingen: J.C.B. Mohr, 1979) pp. 49–62.

AGRICULTURAL ADJUSTMENT ACT OF 1937, 48 Stat. 3, 7 U.S.C. 601 *et seq.*

America Economica (New York: Dow Jones, December 1991).

American Fruit Growers, supra, Hutches v. Renfroe, 200 F.2d 337, 5th Cir. (1952).

American Fruit Growers v. U.S., 105 F. 2d 722, 723–724, 9th Cir. (1939).

ANTOSHAK, ROBERT P., and TERRY L. LEITZELL, 'A Bold Plan: Free Trade', *ATI*, vol. 20, no. 5 (May 1991) pp. 26–28.

AQUINO, A., 'Intra-Industry Trade and Intra-Industry Specialization as Concurrent Sources of International Trade in Manufactures', *Weltwirtschaftliches Archiv*, vol. 114 (1978) pp. 275–95.

AQUINO, A., 'The Measurement of Intra-Industry Trade When Overall Trade is Imbalanced', *Weltwirtschaftliches Archiv*, vol. 117 (1981) pp. 763–6.

AUDLEY, JOHN J., 'Critical Analysis of the February, 1992 Draft North American Free Trade Agreement' (Washington DC: Sierra Club Center for Environmental Innovation, 1992(a)).

AUDLEY, JOHN J., Testimony Before the House Subcommittee on Trade, 17 September 1992.

AUERBACH, STUART, 'Bilateral Trade Packets Worry Experts', *Washington (DC) Post*, 3 November 1988.

AUERBACH, STUART, 'Splitting Protectionist Seems: Mexican Trade Pact Unravels the Once-Durable Textile Lobby', *The Washington Post*, 12 May 1991, pp. H1 and H8.

AWANOHARA, SUSUMU, 'America's Back Door', *Far Eastern Economic Review*, vol. 153, no. 28 (11 July 1991) pp. 44–6.

AYERS, RONALD M., 'Problems of Human Capital Investment along the U.S.-Mexican Border: A Theoretical Perspective', in Khosrow Fatemi (ed.), *U.S.–Mexican Economic Relations* (New York: Praeger, 1988).

BAKER, S. and L. WALKER, 'The American Dream is Alive and Well – in Mexico', *Business Week*, 30 September 1991, pp. 102, 105.

BALASSA, B., *Economic Development and Integration* (Mexico, DF: Centro de Estudios Monetarios Latino Americanos, 1965).

274

Bibliography 275

BALASSA, B., 'Tariff Reductions and Trade in Manufactures among the Industrial Countries', *American Economic Review*, vol. 56, 1966, pp. 466–73.

BALASSA, B., 'Intra-Industry Trade and the Integration of Developing Countries in the World Economy', in H. Giersch (ed.), *On the Economics of Intra-Industry Trade* (Tübingen: J.C.B. Mohr, 1979) pp. 245–270.

BALDWIN, JOHN and PAUL K. GORECKI, 'The Relationship Between Trade and Tariff Patterns and the Efficiency of the Canadian Manufacturing Sector in the 1970s: A Summary', in John Whalley with Roderick Hill (eds), *Canada–United States Free Trade* (Toronto: University of Toronto, 1985).

BARRETT, JOYCE, 'Saludos Amigos; The Push for Free Trade', *Women's Wear Daily*, vol. 162 (17 July 1991) pp. 12–14.

BARRIO, FEDERICO, 'History and Perspectives of the Maquiladora Industry in Mexico', in *In-Bond Industry/Industria Maquiladora* (Mexico, DF: ASI, 1988).

BEACH, DEBRA, 'Mexico Pins Hopes on Free-Trade Pact', *Houston Chronicle*, 26 June 1990.

BERGSTRAND, J., 'Measurement and Determinants of Intra-Industry International Trade', in P.K.M. Tharakan (ed.), *Intra-Industry Trade: Empirical and Methodological Aspects* (Amsterdam: North Holland, 1983) pp. 201–53.

BERMUDEZ, SERGIO, 'Linking Maquiladoras to Local Industry', in Richard L. Bolin (ed.), *Linking the Export Processing Zone to Local Industry* (Flagstaff, AZ: Flagstaff Institute, 1990(a)), pp. 115–25).

BERMUDEZ, SERGIO, 'Transfer of Technology and Management in Private Export Industrial Parks in Mexico', in Richard L. Bolin (ed.), *Technology Transfer and Management in Export Processing Zones* (Flagstaff AZ: Flagstaff Institute, 1990(b))).

BILATERAL TRADE AGREEMENTS, 101st Congress, 1st Session, 1989, p. 82.

BLACK, SUSAN, 'Speaking Out on Free Trade: What a Mexican Deal Could Mean (Part 1)', *Bobbin*, vol. 32, no. 12 (August 1991(a)) pp. 76–84.

BLACK, SUSAN, 'Speaking Out on Free Trade: What a Mexican Deal Could Mean (Part 2)', *Bobbin*, vol. 33, no. 1 (September 1991(b)) pp. 118–26.

BLACK, SUSAN S., '1991 807/CBI Comparative Analysis', *Bobbin*, vol. 33, no. 3 (November 1991(c)) p. 48A.

BLOMQVIST, AKE G., 'International Migration of Educated Manpower and Social Rates of Return to Education in LDCs', *International Economic Review*, vol. 27, no. 1 (1986) pp. 165–74.

BOLIN, RICHARD L. (ed.), *Production Sharing: A Conference with Peter Drucker* (Flagstaff AZ: Flagstaff Institute, 1988).

BOLIN, RICHARD L., 'Maquiladora History and Prospects', *Journal of the Flagstaff Institute* (March 1991) pp. 53–6.

BORDé, F. and P. CROSS, 'Tariffs in Canada-U.S. Trade', *Canadian Economic Observer* (Ottawa: Statistics Canada, 1989) pp. 3.1–3.5.

BORJAS, GEORGE, RICHARD FREEMAN and LAWRENCE KATZ, *On the Labor Market Effects of Migration and Trade*, NBER working paper 3761 (Cambridge: NBER, June 1991).

Bramsen v. Hardin, 346 F. Supp. 934, Fla. 1972 (affd on appeal 485 F.2d 112).

BRANNON, J.T. and G.W. LUCKER, 'The Impact of Mexico's Economic Crisis on the Demographic Composition of the Maquiladora Labor Force', report prepared for the U.S. Department of Labor (August 1988).

BUENO, GERARDO and BARRY W. POULSON, 'Trade and Development Policy in Mexico in the Context of North American Relations', *North American Review of Economics and Finance* (Fall 1990).

BUREAU OF LABOR STATISTICS, 'Current Labor Statistics: International Comparisons Data', *Monthly Labor Review*, vol. 113 (November 1990) pp. 98–100.

BUSH, GEORGE, 'Response to Congressional Concerns Arising from the NAFTA', 1 May 1991(a).

BUSH, GEORGE, 'Response of the Administration to Issues Raised in Connection with the Negotiation of a North American Free Trade Agreement', transmitted to the Congress, 1 May 1991(b).

BUSINESS INTERNATIONAL CORP., 'How to Use Mexico's In-Bond Industry', *Handbooks for Latin America* (1986).

CABEZA RESENDEZ, CARLOS E.Z., 'Theory and Practice of Trade Liberalization', doctoral dissertation in economics (University of Texas at Austin, 1991).

CAMERON, DAVID R., 'The 1992 Initiative: Causes and Consequences', in Alberta M. Sbragia (ed.), *Euro-Politics: Institutions and Policymaking in the 'New' European Community* (Washington, DC: The Brookings Institution, 1992) pp. 23–74.

CAMPBELL, BRUCE, 'Restructuring the Economy: Canada into the Free Trade Era', in Ricardo Grinspun and Maxwell A. Cameron (eds), *The Political Economy of North American Free Trade* (New York: St. Martin's Press, 1993.

CAVES, R., 'Intra-Industry Trade and Market Structure in the Industrial Countries', *Oxford Economic Papers*, vol. 33 (1981) pp. 203–23.

CENTER FOR INTERNATIONAL ENVIRONMENTAL LAW, 'Preliminary Overview of Environmental Concerns Arising from the NAFTA' (August 1992).

CHARNOVITZ, STEVE, 'GATT and the Environment: Examining the Issues', *International Environmental Affairs*, vol. 203 (1992).

CHRISPIN, BARBARA R., 'Employment and Manpower Development in the Maquiladora Industry: Reaching Maturity', in Khosrow Fatemi (ed.), *The Maquiladora Industry: Economic Solution or Problem?* (New York: Praeger, 1990) pp. 71–88.

CHRISTENSEN, ERIC and SAMANTHA GEFFIN, 'GATT Sets its Net on Environmental Regulation: The GATT Panel Ruling on Mexican Yellowfin Tuna Imports and the Need for Reform of the International Trading System', *Miami Inter-American Legal Review*, vol. 569 (1991–2).

CLEMENT, N.C. and S.B. JENNER, 'Location Decisions Regarding Maquiladora/In-Bond Plants Operating in Baja California', Mexico Border Issue Series no. 3 (San Diego, Calif: San Diego State University Institute for Regional Studies of the Californias, 1987).

CLEMENT, NORRIS C. and STEPHEN B. JENNER, 'Mexico's Maquiladora Industry and California's Economy', *Southwest Journal of Business and Economy*, vol. 5, no. 4 (Summer 1988) pp. 1–17.

CLINTON, BILL, 'Expanding Trade and Creating American Jobs', remarks at North Carolina State University, Raleigh, 4 October 1992.

'Clinton Moves in Name of Free Trade may be Inviting Deterioration Instead', *The Wall Street Journal*, 5 February 1993, p. A8.

'Clinton reaffirms Support for NAFTA', *Dow Jones News Services*, 8 January 1993.

CODE OF FEDERAL REGISTER, Volume 54 C.F.R. 220, 1989.

CODE OF FEDERAL REGISTER, Volume 55 C.F.R. 13675, 9515, April 19, 1990.

CODE OF FEDERAL REGISTER, Volume 56 C.F.R. 5841, Feb. 13, 1991.

COMMISSION FOR THE STUDY OF INTERNATIONAL MIGRATION AND COOPERATIVE ECONOMIC DEVELOPMENT, *Unauthorized Migration: An Economic Development Response* (Washington, DC: U.S. Government Printing Office, 1990).

COMMISSION ON THE SKILLS OF THE AMERICAN WORKFORCE, *America's Choice: High Skills or Low Wages!* (Rochester, NY: National Center on Education and the Economy, 1990).

COMMITTEE FOR THE PROMOTION OF INVESTMENT IN MEXICO, *Mexico: Economic and Business Overview* (Mexico City: Committee for the Promotion of Investment in Mexico, June 1990).

'Congress and Mexico, Bordering on Change: A Report of the CSIS Congressional Study Group on Mexico', vol. 11 (Washington, DC: Center For Strategic and International Studies, 1989).

CORNELL, BRADFORD and ALAN C. SHAPIRO, 'Managing Foreign Exchange Risks', in Joel M. Stern and Donald H. Chew Jr (eds), *New Developments in International Finance* (New York: Basil Blackwell, 1988) pp. 44–59.

COSTANZA, ROBERT (ed.), *Ecological Economics* (New York: Columbia University Press, 1991).

COUGHLIN, CLETUS C., 'What do Economic Models Tell Us About the Effects of the U.S.–Canada Free Trade Agreement?', *Review of the Federal Reserve Bank of St. Louis* (September/October 1990) pp. 40–58.

COUGHLIN, CLETUS C., 'U.S. Trade-Remedy Laws: Do They Facilitate or Hinder Free Trade?', *Review of the Federal Reserve Bank of St. Louis*, vol. 73, no. 4 (July/August 1991) pp. 3–18.

COX, D. and R. HARRIS, 'Trade Liberalization and Industrial Organization: Some Estimates for Canada', *Journal of Political Economy*, vol. 93 (1985) pp. 115–45.

CRAIG, JANE and ANN DUPONT, 'Sourcing Beyond Mexican Hurdles', *Bobbin*, vol. 30, no. 10 (June 1989) pp. 107–12.

CROOKELL, HAROLD, *Canadian–American Trade and Investment Under the Free Trade Agreement* (New York: Quorum Books, 1990) pp. 5–7.

C-SPAN, hearings by the House Agricultural Committee on U.S.–Mexico Free Trade by Secretary of Agriculture Edward Madigan, Secretary of Labor Lynn Martin, Ambassador Carla Hills, U.S. Trade Representative, and Linda Fisher, EPA Office of Pesticide, 23 April 1991.

CULEM, C. and L. LUNDBERG, 'The Product Pattern of Intra-Industry Trade: Stability among Countries over Time', *Weltwirtschaftliches Archiv*, vol. 122 (1986) pp. 113–30.

DACHER, PAUL, *Marketing in Mexico* (Washington, DC: U.S. Department of Commerce, International Trade Administration: Supt. of Docs., United States Government Printing Office, August 1990).

Dallas Times Herald, Employment Section, 14 April 1991.

DAVIS, MIKE, 'U.S.–Mexico Seek Free Trade Accord', *Houston Post*, 28 March 1990.

DAWSON, BILL, 'Mexican Market Offers New Opportunities For U.S. Manufacturers of Yarn and Fabric', *Business America*, vol. 111, no. 12 (18 June 1990) p. 23.

DAWSON, BILL, 'Mexican Market Continues to Offer Opportunities for U.S. Manufacturers of Yarn and Fabric', *Business America*, vol. 112, no. 6 (25 March 1991) p. 32.

D'CRUZ, JOSEPH R. and ALAN M. RUGMAN, *New Compacts for Canadian Competitiveness* (Toronto: Kodak Canada, 1992).

DEFOREST, MARIAH E., 'Are Maquiladoras a Menace to U.S. Workers?' *Business Horizons* (November–December 1991) p. 83.

DENNY, MICHAEL and MELVYN FUSS, 'Productivity: A Selective Survey of Recent Developments and the Canadian Experience', discussion paper series (Toronto: Ontario Economic Council, 1982).

DEPARTMENT OF STATE, 'Mexico's Economic Reforms', mimeo (1991).

DESAI, MEGHNAD and WILLIAM LOW, 'Measuring the Opportunity for Product Innovation', in Marcello De Cecco (ed.), *Changing Money: Financial Innovation in Developed Countries* (New York: Basil Blackwell, 1987) pp. 112–40.

DESTLER, I.M., *American Trade Politics*, second edition (Washington, DC: Institute for International Economics, 1992).

DOERN, G. BRUCE and BRIAN W. TOMLIN, *Faith and Fear: The Free Trade Story* (Toronto: Stoddart, 1991).

DORNBUSCH, RUDIGER, 'If Mexico Prospers, So Will We', *Wall Street Journal*, 10 April 1991.

DOWD, A.R., 'Exchange of Letters on Issues Concerning the Negotiation of a North American Free Trade Agreement', Committee on Ways and Means, U.S. House of Representatives, 102D Congress, First Session, 1 May 1991.

DOWD, A.R., 'Viva Free Trade with Mexico', *Fortune*, vol. 123, no. 13 (17 June 1991) pp. 97, 100.

DRABEK, Z. and D. GREENAWAY, 'Economic Integration and Intra-Industry Trade: The EEC and CMEA Compared', *Kyklos*, vol. 37 (1984) pp. 444–69.

DRUCKER, PETER, 'Mexico's Ugly Duckling – the Maquiladora', *Wall Street Journal*, 4 October 1990, p. A20.

DUNKEL, ARTHUR, *Trade and the Environment* (Geneva, Switzerland: General Agreement on Tariffs and Trade, 1992).

ECONOMIC COMMISSION FOR EUROPE, *Economic Survey of Europe in 1987–88* (New York: United Nations, 1988).

ECONOMIC COMMISSION FOR EUROPE, *Economic Survey of Europe in 1988–89* (New York: United Nations, 1989).

EITEMAN, DAVID, ARTHUR STONEHILL and MICHAEL MOFFETT, *Multinational Business Finance*, sixth edition (Reading, MA: Addison-Wesley, 1992).

ELLISON, KATHERINE, 'Eyes on the Future, Mexico Changes Rapidly', *The Charlotte Observer*, 31 March 1991, p. 13A.

'Enormously Consequential', *Forbes* (April 1980).

ENVIRONMENTAL PROTECTION AGENCY, *Integrated Environmental Program for the Mexican–U.S. Border Area*, first stage (February 1992).

ERB, GUY F., Statement Before the Subcommittee on International Trade, Committee on Finance, United States Senate, 13 March 1989.

ETHIER, WILFRED J., *Modern International Economics*, second edition (New York: Norton, 1988).

FATEMI, KHOSROW, 'The Undocumented Immigrant: A Socioeconomic Profile', *Journal of Borderland Studies*, vol. II, no. 2 (Fall 1987).

FATEMI, KHOSROW (ed.), *U.S.–Mexican Economic Relations: Prospects and Problems* (New York: Praeger, 1988).

FATEMI, KHOSROW, 'Introduction', in Khosrow Fatemi (ed.), *The Maquiladora Industry: Economic Solution or Problem?* (New York: Praeger, 1990).

FATEMI, KHOSROW and JIM GIERMANSKI, 'The Maquiladora Industry', paper presented at the Association for Global Business, Orlando, FL (November 1990).

FAUX, JEFF and THEA LEE, 'The Effect of George Bush's NAFTA on American Workers: Ladder Up or Ladder Down?' (Washington, DC: Economic Policy Institute, 1992).

FIDLER, STEPHEN, 'Mexico's Pulling Power', *Financial Times*, 21 July 1992, p. 4.

FINGER, J., 'Trade Overlap and Intra-Industry Trade', *Economic Inquiry*, vol. 13 (1975) pp. 581–9.

FINGER J.M. and M.E. KREINEN, 'A Measure of Export Similarity and its Possible Uses', *The Economic Journal* (December 1979).

FINNERTY, JOHN D., 'Financial Engineering in Corporate Finance: An Overview', *Financial Management*, vol. 17, no. 4 (Winter 1988) pp. 14–33.

FLANAGAN, ROBERT J. *et al.*, *Economics of the Employment Relationship* (Glenview, IL: Scott, Foresman and Company, 1989).

FOGARASI, J., I. MOHN and L. SUBRIN, 'Exporting in North America? The Commerce Department Can Help', *Business America*, vol. 12, no. 7 (8 April 1991) pp. 36–9.

'Free Trade Pact Talks with Chile to Await North American Pact', *The Wall Street Journal*, 14 May 1992, p. A2.

FULLER, KATHRYN S., Testimony Before the Senate Subcommittee on International Trade, 16 September 1992.

GALVEZ, S., 'North American Free Trade Agreement: A Mexican perspective', *Financier*, vol. 15, no. 5 (May 1991) pp. 23–5.

GAMBRILL, MONICA-CLAIRE, *La Politica Salarial de las maquiladoras versus la Industria Nacional* (Mexico City: Centro de Estudios sobre los Estados Unidos de Norteamerica, Universidad Nacional Autonoma de Mexico, 1986).

GAROYAN, 'Marketing Order', 23 U.C. Davis L.R. 697–712 (Spring 1990).

GATT, October 1947, T.I.A.S. no. 1700, 55 U.N.T.S.S. 184, 1947.

GATT, 'Uruguay Round of the General Agreement on Tariffs and Trade', 12 December 1992.

GATT SECRETARIAT, 'Trade and the Environment' (Geneva: GATT, February 1992).

GEORGE, EDWARD Y., 'What Does the Future Hold for the Maquiladora Industry?', in Khosrow Fatemi (ed.), *The Maquiladora Industry: Economic Solution or Problem?* (New York: Praeger, 1990) pp. 219–33.

GEPHARDT, RICHARD, Remarks of Congressman Richard Gephardt: Address on the Status of the North American Free Trade Agreement Before the Institute for International Economics, 27 July 1992.

GIERMANSKI, JIM, KELLY S. KIRKLAND, EDUARDO MARTINEZ, DAVID M. NEIPERT and TOM TETZEL, *U.S. Trucking in Mexico: A Free Trade Issue* (Laredo, TX: Laredo State University, September 1990).

GIERSCH, H. (ed.), *The International Division of Labour* (Tübingen: J.C.B. Mohr, 1974).

GIERSCH, H. (ed.), *On the Economics of Intra-Industry Trade* (Tübingen: J.C.B. Mohr, 1979).

GILBREATH, KENT, 'A Businessman's Guide to the Mexican Economy', *Columbia Journal of World Business*, vol. 21, no. 2 (Summer 1986) pp. 3–14.

GLEJSER, H., K. GOOSSENS and M. VANDEN EEDE, 'Inter-Industry versus Intra-Industry Specialization in Exports and Imports (1959–1970–1973)', *Journal of International Economics*, vol. 12 (1982) pp. 363–9.

GLOBERMAN, STEVEN, *Continental Accord: North American Economic Integration* (Toronto: The Fraser Institute, 1991).

GOODLAND, ROBERT, HERMAN DALY and SALAH EL SERAFY, *Environmentally Sustainable Economic Development: Building on Brundtland* (The World Bank, Environment Department Working Paper No. 46, 1991).

GOVERNMENT OF CANADA, MINISTRY OF FINANCE, *The Canada-US Free Trade Agreement: An Economic Assessment* (Ottawa: Minister of Supply and Services, 1988).

GRAY, H.P., 'Intra-Industry Trade: The Effects of Different Levels of Data Aggregation', in H. Giersch (ed.), *On the Economics of Intra-Industry Trade* (Tübingen: J.C.B. Mohr, 1979) pp. 87–110.

GRAYSON, GEORGE W., 'Mexico's New Politics', *Commonweal*, 25 October 1991.

GREENAWAY, D., 'Intra-Industry Trade, Intra-Firm Trade and European Integration: Evidence, Gains and Policy Aspects', *Journal of Common Market Studies*, vol. 26, 1987, pp. 153–72.

GREENAWAY, D. and C. MILNER, 'Trade Imbalance Effects in the Measurement of Intra-Industry Trade', *Weltwirtschaftliches Archiv*, vol. 117 (1981) pp. 756–62.

GREENAWAY, D. and C. MILNER, 'On the Measurement of Intra-Industry Trade', *The Economic Journal*, vol. 93 (1983) pp. 900–8.

GREENAWAY D. and C. MILNER, 'Intra-Industry Trade: Current Perspectives and Unresolved Issues', *Weltwirtschaftliches Archiv*, vol. 123 (1987) pp. 39–57.

GRINSPUN, RICARDO and MAXWELL A. CAMERON (eds), *The Political Economy of North American Free Trade* (New York: St. Martin's Press, 1993).

GROVES, C., 'Free Trade Agreement – An Opportunity for the Work Force of the Year 2000', presented at teh Central Power and Light Management Conference, December 1990.

GRUBEL, H., 'Intra-Industry Specialization and the Pattern of Trade', *Canadian Journal of Economics and Political Science*, vol. 33 (1967) pp. 374–88.

GRUBEL, H. and P. LLOYD, 'The Empirical Measurement of Intra-Industry Trade', *The Economic Record*, vol. 47 (1971) pp. 494–517.

GRUBEL, H. and P. LLOYD, *Intra-Industry Trade: The Theory and Measurement of International Trade in Differentiated Products* (New York, NY: John Wiley & Sons, 1975).

GUNN, ERIK, 'Canada Still Leery of Free-Trade Pact', *Milwaukee Journal*, 18 March 1990.

GRUNWALD, JOSEPH, 'Prepared Remarks', in Richard L. Bolin (ed.), *Production Sharing: A Conference with Peter Drucker* (Flagstaff AZ: Flagstaff Institute, 1988).

HANSEN, NILES, *The Border Economy: Regional Development in the Southwest* (Austin, TX: University of Texas Press, 1981).

HARDIN, GARRETT, 'The Tragedy of the Commons', *Science*, vol. 162 (December 1968) pp. 1243–8.

HARRIS, RICHARD, *Trade, Industrial Policy and International Competition* (Toronto: University of Toronto Press, 1985(a)).

HARRIS, RICHARD, 'Summary of a Project on the General Equilibrium Evaluation of Canadian Trade Policy', in John Whalley with Roderick Hill (eds), *Canada-United States Free Trade* (Toronto: University of Toronto, 1985(b)).

HARRIS, RICHARD with DAVID COX, *Trade, Industrial Policy, and Canadian Manufacturing* (Toronto: Ontario Economic Council, 1983).

HART, MICHAEL, *A North American Free Trade Agreement: Strategic Implications for Canada* (Ottawa: Centre for Trade Policy and Law, 1990).

HART, MICHAEL, 'A Canadian Perspective on the 1987 Canada-United States Free Trade Agreement', in Glen E. Lich and Joseph A. McKinney (eds), *Region North America: Canada, the United States and Mexico* (Waco, TX: Baylor University Program in Regional Studies, 1990).

HARVEY, A.C., *Time Series Models* (Oxford: Philip Allan, 1981).

HAVRYLYSHYN, O. and E. CIVAN, 'Intra-Industry Trade and the Stage of Development: A Regression Analysis of Industrial and Developing Countries', in P.K.M. Tharakan (ed.), *Intra-Industry Trade: Empirical and Methodological Aspects* (Amsterdam: North Holland, 1983) pp. 111–40.

HAYWOOD, ROBERT C. and RICHARD L. BOLIN, *Sonora 2010*, Flagstaff AZ: Flagstaff Institute, 1989.

HERITAGE FOUNDATION, 'Guidelines for U.S. Negotiations at the Trade Talks with Mexico', *Backgrounder*, no. 861, 18 October 1991.

HILL, RODERICK and JOHN WHALLEY, 'A Possible Canada–U.S. Free Trade Arrangement: A Summary of the Proceedings of a Research Symposium', in John Whalley with Roderick Hill (eds), *Canada–United States Free Trade* (Toronto: University of Toronto, 1985).

HINE, ROBERT, 'Customs Union Enlargement and Adjustment: Spain's Accession to the European Community', *Journal of Common Market Studies*, vol. 28 (September 1989) pp. 1–27.

HINOJOSA-OJEDA, RAUL and SHERMAN ROBINSON, 'Labor Issues in a North American Free Trade Area', in Barry Bosworth and Robert Lawrence, *North American Free Trade: Assessing the Impact* (Washington, DC: The Brookings Institution, 1992).

HOWARD, R., 'The Designer Organization: Italy's GFT Goes Global', *Harvard Business Review*, vol. 69, no. 5 (September/October 1991) pp. 28–44.

HU, SHENG CHENG, 'Education and Economic Growth', *Review of Economic Studies*, vol. 43 (1976) pp. 509–19.

HUDSON, STEWART J., Testimony Before the Senate Subcommittee on International Trade, 16 September 1992.

HUFBAUER, G. and J. CHILAS, 'Specialization by Industrial Countries: Extent and Consequences', in H. Giersch (ed.), *The International Division of Labour* (Tübingen: J.C.B. Mohr, 1974) pp. 3–38.

HUFBAUER, GARY C. and JEFFREY J. SCHOTT, *North American Free Trade: Issues and Recommendations* (Washington, DC: Institute for International Economics, 1992).

HUGHES, R., Remarks before the U.S.–Mexico Chamber of Commerce, Southwest Chapter, 5 February 1991.

IBARRA, MARCO A. and GUILLERMO SANDER, 'Maquiladora Compensation and Benefits on the Mexican Border', in *In-Bond Industry/Industria Maquiladora* (Mexico, DF: ASI, 1988) pp. 59–66.

ILO/UNCTC, *Economic and Social Effects of Multinational Enterprises in Export Processing Zones* (Geneva: International Labour Office, 1988).

INDUSTRY, SCIENCE, TECHNOLOGY CANADA, *The North American Free Trade Agreement*, Business and Professional Services Directorate, Services and Construction Industries Branch, 4 September 1991.

INFORUM-CIMAT, *Industrial Effects of a Free Trade Agreement Between Mexico and the USA* (Springfield, VA: National Technical Information Service, 1990).

INTERNATIONAL MONETARY FUND, *Direction of Trade Statistics Yearbook 1991 and 1992* (Washington, DC: International Monetary Fund, 1992).

INTERNATIONAL MONETARY FUND, *International Financial Statistics* (Washington, DC: International Monetary Fund, 1992).

JACQUE, LAURENT L., 'Management of Foreign Exchange Risk: A Review Article', *Journal of International Business Studies*, vol. 12, no. 1 (Spring/Summer 1981) pp. 81–101.

JESSWEIN, KURT R., 'Adoption Criteria for New Foreign Exchange Risk Management Products: Some Implications for Financial Innovation Theory', dissertation in development (University of South Carolina, 1992).

KEELEY, TERRENCE, 'Financial Innovation and Social Benefit', in John Calverly and Richard O'Brian, (eds), *Finance and the International Economy* (Oxford: Oxford University Press, 1987) pp. 117–35.

KIM, SUNWOONG, 'Labor Specialization and the Extent of the Market', *Journal of Political Economy*, vol. 97, no. 3 (1989) pp. 692–705.

KINDLEBERGER, CHARLES P., *Manias, Panics, and Crashes* (New York: Basic Books, 1978).

KING, LARRY, transcripts of and interview with President Bush, 4 October 1992.

KIRKLAND, LANE, 'U.S.–Mexico Trade Pact: A Disaster Worthy of Stalin's Worst', *Wall Street Journal*, 18 April 1991.

KOL, J. and R. RAYMENT, 'Allyn Young Specialization and Intermediate Goods in Intra-Industry Trade', in P.K.M. Tharakan and J. Kol (eds), *Intra-Industry Trade: Empirical and Methodological Aspects* (Amsterdam: North Holland, 1989) pp. 51–68.

KREYE, OTTO, JUERGEN HEINRICHS and FOLKER FROBEL, 'Export Processing Zones in Developing Countries', ILO Multinational Enterprises Programme, Working Paper no. 43 (Geneva: International Labour Office, 1987).

KRUGMAN, P., 'Intra-Industry Specialization and the Gains from Trade', *Journal of Political Economy*, vol. 89 (1981) pp. 959–73.

KRUGMAN, PAUL, 'Macroeconomic Adjustment and Entry into the EC: A Note', in Christopher Bliss and Jorge De Macedo, *Unity with Diversity in the European Economy: the Community's Southern Frontier* (Cambridge: Cambridge University Press, 1990).

LAZAR, FRED, *The New Protectionism: Non-Tariff Barriers and Their Effects on Canada* (Toronto: Lorimer, 1981).

LEVHARI, DAVID and YORAM WEISS, 'The Effect of Risk on the Investment in Human Capital', *The American Economic Review*, vol. 64 (1974) pp. 950–63.

LINDQUIST, DIANE, 'Mexican Leader Pushing Even Harder for Free Trade', *San Diego Union*, 6 September 1990.

LINDSEY, LAWRENCE B., 'America's Growing Economic Lead', *The Wall Street Journal*, 7 February 1992, p. A14.

LIPIETZ, ALAIN, 'Le Kaleidoscope des "sud"', in Robert Boyer, *Capitalismes fin de Siecle* (Paris: PUF, 1986) pp. 203–24.

LUCKER, G.W. and A. ALVAREZ, 'Exploitation or Exaggeration: A Worker's Eye View of Maquiladora Work', *Southwest Journal of Business and Economics*, Summer 1984, pp. 11–18.

LUCKER, G. W. and A. ALVAREZ, 'Controlling Maquiladora Turnover Through Personnel Selection', *Southwest Journal of Business and Economics*, vol. 2, no. 3 (Spring 1985) pp. 1–10.

LUSTIG, NORA, BARRY P. BOSWORTH and ROBERT Z. LAWRENCE (eds.), *North American Free Trade: Assessing the Impact* (Washington, DC, Brookings Institution, 1992).

MACDONALD REPORT, *Report: Royal Commission on the Economic Union and Development Prospects for Canada*, Volume I (Ottawa: Minister of Supply and Services, 1985).

MADDISON, ANGUS, 'Growth and Slowdown in Advanced Capitalist Economies', *Journal of Economic Literature*, vol. 25 (June 1987) pp. 649–98.

MAGNIER, MARK, 'Clouds Hang Over Mexico's Twin-Plant Success', *Journal of Commerce and Commercial*, vol. 379, (1 March 1989) pp. 1A, 3A.

MANNING, R., 'Optimal Aggregative Development of a Skilled Workforce', *Quarterly Journal of Economics*, vol. 89 (1975) pp. 504–11.

MANNING, R., 'Issues in Optimal Educational Policy in the Context of Balanced Growth', *Journal of Economic Theory*, vol. 13 (1976) pp. 380–95.

MANNING, R., 'Two Theorems Concerning Optimal Education Policy in Balanced Growth', *International Economic Review*, vol. 23 (1982) pp. 83–106.

MARCH, B., 'Mexico: On the Fast Track', *The IBD*, Inter-American Development Bank (October 1991) pp. 4, 5.

MARSH, B., 'Mexico Becomes More Neighborly', *Bobbin* (June 1991) pp. 48–51.

MARTINEZ, OSCAR J., 'The Foreign Orientation of the Mexican Border Economy', *Border Perspectives*, no. 2 (May 1983).

MCALLISTER, EUGENE J., 'NAFTA: A Critical Priority', *US Department of State Dispatch*, vol. 2 (6 May 1991) pp. 322–5.

MENDES, MARQUES, 'The Contribution of the European Community to Economic Growth', *Journal of Common Market Studies*, vol. 24 (June 1986).

MERIDA, KEVIN, 'U.S., Mexico to Explore Agreement on Free Trade', *Dallas Morning News*, 12 June 1990.

Mexico 2000: A Classical Analysis of the Mexican Economy and the Case for Supply-Side Economic Reforms (New Jersey: Polyconomics, Inc., 1990).

MICHEL, RICHARD H., 'General Electric Company and Production Sharing', in Richard L. Bolin (ed.), *Production Sharing: A Conference with Peter Drucker* (Flagstaff, AZ: Flagstaff Institute, 1988) p. 11.

MILLS, TERENCE C., *Time Series Techniques for Economists* (Cambridge: Cambridge University Press, 1990).

MORICI, PETER, *Trade Talks with Mexico: A Time for Realism* (Washington, DC: National Planning Association, 1991).

MORTON, PETER, 'Not Always Good Neighbors', *Financial Post*, 16 March 1991.

NATURAL RESOURCES DEFENSE COUNCIL, 'Environmental Safeguards for the North American Free Trade Agreement', June 1992.

NAZARIO, SONIA, 'Boom and Dispair', *Wall Street Journal*, 22 September 1989.

North American Free Trade Agreement (Washington, DC: U.S. Special Trade Representative's Office, 1992).

NOWICKI, LAWRENCE W., 'Economic Recession and New Investment Strategies at the International Level' (Paris: OECD Development Center, 1982).

OECD, *OECD Economic Surveys: Spain* (Paris: OECD, 1990).

OECD, *SOPEMI 1989* (Paris: OECD, 1990).

OECD, *OECD Economic Surveys: Greece* (Paris: OECD, 1990).

OECD, *OECD Economic Surveys: Portugal* (Paris: OECD, 1992).

OFFICE OF THE U.S. TRADE REPRESENTATIVE, 'Report of the Administration on the North American Free Trade Agreement and Actions Taken in Fulfillment of the May 1, 1991 Commitments' (Washington, DC, 1992).

'Overview of United States–Mexico Relations', 101st Congress, 1st Session (1989) pp. 5, 45, 46.

PATRICK, J. MICHAEL, 'The Employment Impact of Maquiladoras Along the U.S. Border', in Khosrow Fatemi (ed.), *The Maquiladora Industry: Economic Solution or Problem?* (New York: Praeger, 1990) pp. 57–70.

P C GLOBE (database) (Tempe, AZ: P.C. Globe, Inc., 1987, 1988, 1989).
PETERSON, R.T., 'Mexico: A Market, Not Just a Place to Produce', *Marketing News*, vol. 25, no. 12 (10 June 1991).
POLLUTION PROBE-CANADA/NATIONAL WILDLIFE FEDERA-TION, 'Binational Statement of Environmental Safeguards that Should be Included in the NAFTA' (May 1992).
POULSON, BARRY W., 'A North American Free Trade Area', paper presented at the North American Economics and Finance Association Meeting, Mexico City, August 1990.
POWER, PAUL, 'Farmers Object to Free Trade with Mexico', *Tampa Tribune*, 27 September 1990.
'Proposed Negotiations for Free Trade Agreement with Mexico', hearings before the Subcommittee on Trade of the House of Representatives Committee on Ways and Means, 102nd Congress, 1st Session (1991).
'Proposed United States–Mexico Free Trade Agreement and Fast Track Authority Hearing', before the Committee on Agriculture of the House of Representatives Committee on Agriculture', 102nd Congress, 1st Session (1991).
PURCELL, JOHN H. and DIRK DAMRAU, *Mexico: A World Class Economy in the 1990s* (New York: Salomon Brothers Sovereign Assessment Group, 1990).
RANGAZAS, PETER, 'Human Capital Investment in Wealth-Constrained Families with Two-Sided Altruism', *Economics Letters*, vol. 35 (1991) pp. 137–41.
RANSON, DAVID, 'GNP Always Outwits Monetary Policy', *Wall Street Journal*, 14 April 1991.
RAO, P. SOMESHWAR and TONY LEMPRIÈRE, *An Analysis of the Linkages Between Canadian Trade Flows, Productivity, and Costs*, working paper no. 46 (Ottawa: Economic Council of Canada, 1992(a)).
RAO, P. SOMESHWAR and TONY LEMPRIÈRE, 'A Comparison of the Total Factor Productivity and Total Cost Performance of Canadian and U.S. Industries', working paper no. 45 (Ottawa: Economic Council of Canada, 1992(b)).
RAWORTH, PHILIP, 'A North American Common Market: A Canadian Perspective', in Khosrow Fatemi (ed.), *International Trade and Finance: A North American Perspective* (New York: Praeger Publishers, 1988, pp. 270–84).
RAZIN, ASSAF, 'Optimum Investment in Human Capital', *Review of Economic Studies*, vol. 39 (1972) pp. 455–60.
REILLY, WILLIAM K., 'Mexico's Environment Will Improve With Free Trade', *Wall Street Journal*, 19 April 1991, p. A15.
REILLY, WILLIAM, testimony before the Senate Subcommittee on International Trade, 16 September 1992.
REYNOLDS, CLARK, 'A United States View', in Sidney Weintraub (ed.), *Industrial Strategy and Planning in Mexico and the United States* (Boulder, CO: The Westview Press, 1986).
RICH, JAN GILBREATH, *Planning the Border's Future: The Mexican-U.S. Integrated Border Environmental Plan* (Austin, TX: Lyndon B. Johnson School of Public Affairs, U.S.–Mexican Policy Studies Program, 1992).

RICHARD, JOHN D. and RICHARD G. DEARDEN, *The Canada–U.S. Free Trade Agreement: Commentary and Related Documents* (Toronto: Commerce Clearing House-Canadian, 1987).

RILEY, KAREN, 'Free-Trade Pact With Mexico Advocated', *Washington (DC) Times*, 19 October 1989.

RIVAS, FRANCISCO X. and FEDERICO SADA, 'Personnel Turnover in the Maquiladora Industry', in *In-Bond Industry/Industria Maquiladora* (Mexico, DF: ASI, 1988).

ROGOZINSKI, JACQUES, 'Learning the ABCs of Mexico's Privatization Process', *The Wall Street Journal*, 15 May 1992, p. A11.

ROSENZWEIG, MARK R., 'Human Capital, Population Growth, and Economic Development: Beyond Correlation', *Journal of Policy Modeling*, vol. 10, no. 1 (1988) pp. 83–111.

ROSS, STEPHEN A., 'Institutional Markets, Financial Marketing, and Financial Innovation', *Journal of Finance*, vol. 44, no. 3 (July 1989) pp. 541–56.

RUGMAN, ALAN M., 'Business Concerns About Implementing the Free Trade Agreement', *Business Quarterly* (Spring 1989) p. 25.

SAGOFF, MARK, *The Economy of the Earth* (New York: Cambridge University Press, 1988).

SALINAS, CARLOS R., 'The Maquiladoras of Mexico: An Effort to Understand the Controversy', *Southwest Journal of Business and Economics* (Fall 1986) pp. 18–29.

SALTER, STEPHEN B., 'Classification of Financial Reporting Systems and a Test of their Environmental Determinants', dissertation (University of South Carolina, 1991).

SANDERSON, S. and R. HAYES, 'Mexico – Opening Ahead of Eastern Europe', *Harvard Business Review* (September–October 1990) pp. 32–41.

SATCHELL, MICHAEL, 'A "Vicious Circle of Poison": New Questions about American Exports of Powerful Pesticides', *U.S. News and World Reports*, 10 June 1991, pp. 31–2.

SCHEINMAN, MARC N., 'Report on the Present Status of Maquiladoras', in Khosrow Fatemi (ed.), *The Maquiladora Industry: Economic Solution or Problem?* (New York: Praeger, 1990, pp. 19–31).

SCHOEPFLE, GREGORY K., 'Labor Standards in Export Assembly Operations in Mexico and the Carribean', paper presented at the North American Economic and Finance Association, Atlanta, 29 December (Bureau of International Labor Affairs, U.S. Department of Labor, 1989).

SCHOEPFLE, GREGORY K., 'U.S.–Mexico Free Trade Agreement: The "Maquilazation" of Mexico?', Bureau of International Labor Affairs, U.S. Department of Labor (1990).

SCHOEPFLE, GREGORY and PEREZ-LOPEZ, JORGE, 'U.S. Employment Effects of a North American Free Trade Agreement: A Survey of Issues and Estimated Effects', discussion paper 40, Bureau of International Labor Affairs (Washington DC: U.S. Dept. of Labor, 1992).

SCHOTT, JEFFERY J., *More Free Trade Areas* (Washington, DC: Institute for International Economics, May 1989).

SECRETARIA DE COMERCIO Y FOMENTO INDUSTRIAL, *List of Maquiladoras Operating in Mexico*, Xerox, 1988.

'Senator Brock Adams (WA) is concerned about the potato, asparagus, bean and other vegetable crops in the State of Washington', *Spokeman Review*, 24 May 1991, p. A10.

SHAW, KATHRYN L., 'Life-Cycle Labor Supply with Human Capital Accumulation', *International Economic Review*, vol. 30, no. 2 (1989) pp. 431–55.

SIERRA CLUB, *et al.*, 'The Environmental and Consumer Response to the North American Free Trade Agreement', press release, 8 October 1992.

SINCLAIR, SCOTT, 'NAFTA and U.S. Trade Policy: Implications for Canada and Mexico', in Ricardo Grinspun and Maxwell A. Cameron (eds), *The Political Economy of North American Free Trade* (New York: St. Martin's Press, 1993).

SMITH, MURRAY G. and FRANK STONE, (eds), *Assessing the Canada-U.S. Free Trade Agreement* (Halifax: Institute for Research on Public Policy, 1987).

SNOW, ARTHUR and RONALD S. WARREN Jr, 'Human Capital Investment and Labor Supply Under Uncertainty', *International Economic Review*, vol. 31, no. 1 (1990) pp. 195–206.

SOBARZO, HORACIO E., in *Economy-Wide Modeling of the Economic Implications of a FTA with Mexico and a NAFTA with Canada and Mexico* (Washington, DC: United States International Trade Commission, May 1992) pp. 599–652.

SOLANA, F., the Secretary of Foreign Affairs of Mexico, Address Presented at the Dallas-Fort Worth International Trade Conference, 9 May 1991.

SPIERS, J., 'U.S. Trade: Tattered, Not Torn', *Fortune*, 26 August 1991, pp. 21–2.

STARE, METKA, MITRE KOLISEVSKI and MARINKA SMODIS (translator), *Trade Cooperation Among Developing Countries* (Ljubljana, Yugoslavia: Research Center for Cooperation with Developing Countries, 1987).

'Statements of Congressman De Fagio', 137 Cong. Rec. H2478, 24 April 1991.

'Statements of Congressman Smith – Florida', 137 Cong. Rec. H2404, 23 April 1991.

'Statements of Congressman Rahall', 136 Cong. Rec. H2690, 24 May 1991.

Statistical Abstract of the United States 1991 (Washington, DC: U.S. Government Printing Office, 1991).

STATISTICS CANADA, *Aggregate Productivity Measures*, catalogue 15–204E (1992).

STERN, ROBERT M., 'Review of "Trade, Industrial Policy, and Canadian Manufacturing" by Richard G. Harris (with the assistance of David Cox)', *Journal of International Economics*, vol. 19 (1985) pp. 189–200.

STERNTHAL, SUSAN, 'Mexican Fears of Open Trade are Subsiding', *Washington (DC) Times*, 11 June 1990.

STERNTHAL, SUSAN, 'Trade Pact Looking Better to Balkers South of the Border', *Washington (DC) Times*, 18 June 1990.

STODDARD, ELLWYN R., *Maquila-Assembly Plants in Northern Mexico* (The University of Texas at El Paso: Texas Western Press, 1987).

SURO, ROBERTO, 'Border Boom's Dirty Residue Imperils U.S.–Mexico Trade', *New York Times*, 31 March 1991.

SWOBODA, FRANK and MARTHA H. HAMILTON, 'How Virginia Lost Jobs to Texas, Mexico: Closing of AT&T Plant Suggest Complexity of Free Trade Issue', *The Washington Post*, 5 May 1991, pp. H1, H6.

Tariff Act of 1930, 19 U.S.C. 1331(b).

THARAKAN, P.K.M. (ed.), *Intra-Industry Trade: Empirical and Methodological Aspects* (Amsterdam: North Holland, 1983).

THARAKAN, P.K.M. and J. KOL, (eds), *Intra-Industry Trade: Theory, Evidence and Extensions* (New York, NY: St. Martin's Press, 1989).

THUERMER, KAREN E., 'Global Trade', *Global Trade*, vol. 3, no. 2 (February 1991).

TIANO, SUSAN B., 'Women's Work and Unemployment in Northern Mexico', in Vicki L. Ruiz and Susan B. Tiano, (eds), *Women on the U.S.-Mexico Border* (Winchester, MA: Allen and Unwin, 1987).

TODARO, MICHAEL P., *Economic Development in the Third World* (New York: Longman, 1989).

TOLAN, SANDY, 'The Border Boom: Hope and Heartbreak', *New York Times Magazine*, 1 July 1990.

Trade Act of 1974, Generalized System of Preference, 19 U.S.C. 2461.

UCHITELLE, LOUIS, 'Mexico's Plan for Industrial Might: High Technology Plants in the North', *New York Times*, 25 September 1990, pp. D1, D7.

'Understanding concerning a Framework for Trade and Investment', United States–Mexico Economic Relations, 100th Congress, 1st Session (1989).

'Understanding on Trade and Investment Relations', 6 Nov. 1987, United States–Mexico, 27 I.L.M. 438 (1988).

'United States–Mexico Economic Relations', 100th Cong. 1st Session (1989).

U.S. BUREAU OF CENSUS, *1990 Census of Population and Housing*, Public Law 94–171 Data (Washington, DC: U.S. Government Printing Office, 1991).

U.S. CONGRESS, Office of Technology Assessment, *U.S.-Mexico Trade: Pulling Together or Pulling Apart?*, ITE-545 (Washington, DC: U.S. Government Printing Office, October 1992).

'U.S. Countervailing Duty Laws Applied to Mexico', *George Washington Journal of International Law and Economics*, vol. 183 (1984).

U.S. DEPARTMENT OF COMMERCE, American Embassy, *Foreign Economic Trends and their Implications for the U.S.*, vol. 19 (Washington, DC: Government Printing Office, February 1989).

U.S. DEPARTMENT OF COMMERCE, 'Marketing in Mexico', *Oversees Business Report*, OBR 90–09 (1990).

U.S. DEPARTMENT OF LABOR, *Bureau of Labor Statistics Report 817* (1991).

'U.S., EC Agree to Press Japan on Trade Issues', *The Wall Street Journal*, 12 February 1993, p. A2.

U.S. EMBASSY, Mexico, *Current Economic Trends Report* (August 1991).

U.S. GENERAL ACCOUNTING OFFICE, *U.S.-Mexico Trade. Some U.S. Wood Furniture Firms Relocated From Los Angeles Area to Mexico* (Washington, DC: General Accounting Office, 1991).

U.S. INTERNATIONAL TRADE COMMISSION, *Tariff Schedule of the United States Annotated*, USITC Publication 1610 (Washington, DC: United States Government Printing Office, 1987).

U.S. INTERNATIONAL TRADE COMMISSION, *The Use and Economic Impact of TSUS Items 806.30 and 807.00*, publication no. 2053 (Washington, DC: U.S. International Trade Commission, 1988).

U.S. INTERNATIONAL TRADE COMMISSION, *The Use and Economic Impact of TSUS Items 806.30 and 807.00*, report to the Subcommittee on Ways and Means, United States House of Representatives, on Investigation no. 332–244 under Section 332(b) of the Tariff Act of 1930 (June 1988).

U.S. INTERNATIONAL TRADE COMMISSION, *Harmonized Tariff Schedule of the United States*, USITC Publication 2030 (Washington, DC: United States Government Printing Office, 9 June 1989).

U.S. INTERNATIONAL TRADE COMMISSION, *The Likely Impact on the United States of a Free Trade Agreement with Mexico*, publication no. 2353, U.S. International Trade Commission (Washington, DC, 1991).

U.S. INTERNATIONAL TRADE COMMISSION, *Economy-wide Modeling of the Economic Implications of a FTA with Mexico and a NAFTA with Canada and Mexico*, USITC Publication 2508 (Washington, DC, USITC, 1992).

U.S. INTERNATIONAL TRADE COMMISSION publication 2275, 'Recent Trade and Investment Reforms Undertaken by Mexico and Implications for the United States', *Review of Trade and Investment Liberalization Measures by Mexico and Prospects for Future United States–Mexican Relations, Phase I* (Washington, DC: United States International Trade Commission, April 1990(a)).

U.S. INTERNATIONAL TRADE COMMISSION publication 2326, 'Summary of Views on Prospects for Future United States–Mexican Relations', *Review of Trade and Investment Liberalization Measures by Mexico and Prospects for Future United States–Mexican Relations, Phase I* (Washington, DC: United States International Trade Commission, October 1990(b)).

'U.S. Threatens Trade Action Against the EC', *The Wall Street Journal*, 2 February 1993, p. A2.

VARGAS, LUCINDA, 'Economic and Industry Impacts', CIEMEX-WEFA, prepared remarks at 'Shaping the Canada–U.S.–Mexico Free Trade Area', New York Association of Business Economists Conference, 19 March 1991.

Vaughn-Griffin Packing Company v. Hardin 423 F.2d 1094, 5th Cir. (1970).

VERBIT, GILBERT P., *Trade Agreements For Developing Countries* (New York: Columbia University Press, 1969) pp. 56–82.

VERDOORN, P., 'The Intra-Bloc Trade of Benelux', in E.A.G. Robinson (ed.), *Economic Consequences of the Size of Nations*, proceedings of a conference held by the International Economic Association (London, 1960) pp. 291–329.

VERNON, RAYMOND and LOUIS T. WELLS, Jr, *The Manager in the International Economy* (Englewood Cliffs, NJ: Prentice Hall, 1991).

Wallace v. Hudson-Duncan, Co. 98 F.2d 985, 9th Cir. (1939).

WALLACH, LORI, *Critical Analysis of 'The Final Act Text'* (Washington, DC: Public Citizen, December 1991).

WANG, K.S., 'Linkage Effects in Taiwan EPZs', in Richard L. Bolin (ed.), *Linking the Export Processing Zone to Local Industry* (Flagstaff, AZ: Flagstaff Institute, 1990) pp. 23–8.

WARD, JUSTIN, Testimony Before the Senate Subcommittee on International Trade, 16 September 1992.

WARNER, JUDITH A., 'The Sociological Impact of the Maquiladoras', in Khosrow Fatemi (ed.), *The Maquiladora Industry: Economic Solution or Problem?* (New York: Praeger, 1990) pp. 183–97.

WATHEN, THOMAS, *A Guide to Trade and the Environment* (New York: Environmental Grantmakers Association, 1992).

WATKINS, MEL, 'Reservations Concerning a Free Trade Area', in John Whalley with Roderick Hill, (eds), *Canada–United States Free Trade* (University of Toronto Press, 1985).

WAVERMAN, LEONARD, 'A Canadian Vision of North American Economic Integration', in Steven Globerman (ed.), *Continental Accord: North American Economic Integration* (Toronto: The Fraser Institute, 1991) pp. 31–64.

WEINTRAUB, SIDNEY, *Free Trade between Mexico and The United States?* (Washington, DC: The Brookings Institute, 1984).

WEINTRAUB, SIDNEY, *Mexican Trade Policy and the North American Community*, vol. 10, no. 14 (Washington, DC: Center for Strategic and International Studies, 1988).

WEINTRAB, SIDNEY, *A Marriage of Convenience: Relations between Mexico and the United States* (New York: Oxford University Press, 1990) p. 101.

WEINTRAUB, SIDNEY, 'Industrial Integration Policy: U.S. Perspective', in S. Weintraub, L. Rubio F. and A. Jones, (eds), *U.S.–Mexican Industrial Integration* (Boulder, CO: Westview Press, 1991) pp. 49–62.

WEINTRAUB, SIDNEY, F. RUBIO and A. JONES (eds), *U.S.–Mexican Industrial Integration*, Boulder, CO: Westview Press, 1991.

WEINTRAUB, SIDNEY and JAN GILBREATH, 'The Social Side to Free Trade', in *Canada–U.S. Outlook* (a quarterly publication of the National Planning Association), vol. 3, no. 3 (August 1992) pp. 36–59.

WESTERN MAQUILADORA TRADE ASSOCIATION, membership roster, Xerox (1989).

WHALLEY, JOHN, 'Trade, Industrial Policy, and Canadian Manufacturing' by Richard G. Harris (with the assistance of David Cox): a review article", *Canadian Journal of Economics*, vol. 17, no. 2 (1984) pp. 386–98.

WHALLEY, JOHN with RODERICK HILL (eds), *Canada–United States Free Trade* (University of Toronto Press, 1985).

WHARTON ECONOMETRIC FORECASTING ASSOCIATES, 'The Maquiladora Industry Outlook' (1988).

Wileman Bros. & Elliot, Inc. v. Giannini, 909 F.2d 332, 9th Cir. (1990).

WILKINSON, BRUCE, 'Canada–United States Free Trade: The Current Debate', *International Journal*, vol. 62, no. 1 (1987) pp. 199–218.

WILLIAMS, EDWARD J., 'Turnover and Recruitment in the Maquiladora Industry: Causes and Solutions in the Border Context', report submitted to the Office of International Economic Affairs, Bureau of International Labor Affairs, U.S. Department of Labor, November 1988.

WILLIAMS, JOSEPH T., 'Uncertainty and the Accumulation of Human Capital over the Life Cycle', *Journal of Business*, vol. 52, no. 4 (1979) pp. 521–48.

WILSON, PATRICIA, A., 'The New Maquiladoras: Flexible Production in Low-Wage Regions', in Khosrow Fatemi (ed.), *The Maquiladora Industry: Economic Solution or Problem?* (New York: Praeger, 1990).

WIRTH, DAVID, testimony Before the House Committee on Science, Space and Technology, 30 September 1992

WOLCOTT, SUSAN, *The Impact of Accession with the European Community on Spanish Industry*, U.S. Dept. of Labor, Contract J9K9–0076 (1991).

WOMACK, JAMES P., DANIEL T. JONES and DANIEL ROOS, *The Machine That Changed the World* (New York: Macmillan, 1990).

WONNACOTT, PAUL, *The United States and Canada: The Quest for Free Trade* (Washington, DC: Institute for International Economics, 1987).

WONNACOTT, RONALD J., 'Potential Economic Effects of a Canada–U.S. Free Trade Agreement', in John Whalley with Roderick Hill (eds), *Canada–United States Free Trade* (University of Toronto Press, 1985).

WOOD, ADRIAN, 'How Much Does Trade with the South Affect Workers in the North', *World Bank Research Observer*, vol. 6 (January 1991) pp. 19–36.

WORLD BANK, *The World Bank Atlas* (Washington DC: World Bank, 1989).

WORLD BANK, *World Development Report* (Oxford University Press, 1990, 1991, 1992).

WORLD BANK, *Export Processing Zones*, Policy and Research Series 20 (Washington DC: World Bank, 1992).

WORLD COMMISSION ON ENVIRONMENT AND DEVELOPMENT (WCED), *Our Common Future* (New York: Oxford University Press, 1987).

WU, PEI-TSE, 'US Apparel Makers Split Over Mexican Trade Pact', *Journal of Commerce and Commercial*, vol. 388 (4 April 1991) pp. 1A, 5A.

Index